Martin John Sutton

Permanent and Temporary Pastures

Elibron Classics
www.elibron.com

Elibron Classics series.

ISBN 0-543-69569-7 (paperback)
ISBN 0-543-69568-9 (hardcover)

This Elibron Classics Replica Edition is an unabridged facsimile of the edition published in 1911 by Simpkin, Marshall, Hamilton, Kent & Co., Ltd., London.

PERMANENT AND TEMPORARY PASTURES

PRINTED BY
SPOTTISWOODE AND CO. LTD., LONDON
COLCHESTER AND ETON

PERMANENT AND TEMPORARY PASTURES

BY

MARTIN J. SUTTON

CHEVALIER DE LA LÉGION D'HONNEUR; CHEVALIER DE L'ORDRE DU MÉRITE AGRICOLE
FELLOW OF THE LINNEAN SOCIETY
MEMBER OF THE NATIONAL AGRICULTURAL EXAMINATION BOARD
MEMBER OF THE COUNCIL OF THE ROYAL AGRICULTURAL SOCIETY OF ENGLAND FROM 1883 TO 1904
MEMBER OF THE COUNCIL OF THE BATH AND WEST AND SOUTHERN COUNTIES
AGRICULTURAL SOCIETY FROM 1880
KNIGHT OF GRACE OF THE ORDER OF THE HOSPITAL OF ST. JOHN OF JERUSALEM IN ENGLAND
ETC.

POPULAR EDITION

London
SIMPKIN, MARSHALL, HAMILTON, KENT & CO. LIMITED
1911

THE GOLD MEDAL

OF

THE PARIS EXHIBITION, 1900

WAS AWARDED

FOR THE EDITION OF THIS WORK

CONTAINING THE ILLUSTRATIONS

IN NATURAL COLOURS

PREFACE

By gracious permission the First Edition of this work was dedicated to His Majesty King Edward VII.—at that time His Royal Highness The Prince of Wales. I take this renewed opportunity of recording my deep sense of gratitude for the great kindness then extended to me.

To meet a continuous demand six other editions have since been published.

For the present issue the text has been revised throughout and brought up to date, including the latest available statistics concerning the area under grass.

The importance of Temporary Pastures is increasingly recognised, and I am firmly convinced that the future prosperity of farming is intimately bound up with an extension of the system of Leys, varying from one to four years in duration.

Reading: *February* 1911.

CONTENTS

PAGE

INTRODUCTION 1

THE DRAINAGE OF GRASS LAND 3

CULTURAL PREPARATIONS 11

THE SELECTION OF GRASSES AND CLOVERS 19

TYPICAL PRESCRIPTIONS 23

AGRICULTURAL GRASSES (*Gramineæ*)

AGROSTIS ALBA—VAR. STOLONIFERA (*Fiorin*, or *Creeping Bent Grass*) . . . 24

ALOPECURUS PRATENSIS (*Meadow Foxtail*) 25

AMMOPHILA ARUNDINACEA. PSAMMA ARENARIA (*Sea Reed*, *Marrem*, or *Mat Grass*) 28

ANTHOXANTHUM ODORATUM (*Sweet-scented Vernal*) 29

AVENA FLAVESCENS (*Yellow Oat Grass*) 31

AVENA ELATIOR (*Tall Oat Grass*) 32

BROMUS INERMIS (*Awnless Brome Grass*, or *Hungarian Forage Grass*) . . . 34

BROMUS SCHRÆDERI (*Schræder's Brome Grass*) 34

CYNOSURUS CRISTATUS (*Crested Dogstail*) 35

DACTYLIS GLOMERATA (*Rough Cocksfoot*) 37

ELYMUS ARENARIUS (*Sand*, or *Upright Sea Lyme Grass*) 39

FESTUCA PRATENSIS (*Meadow Fescue*) 40

FESTUCA ELATIOR (*Tall Fescue*) 43

FESTUCA OVINA (*Sheep's Fescue*) 44

FESTUCA OVINA TENUIFOLIA (*Fine-leaved Sheep's Fescue*) 45

FESTUCA DURIUSCULA (*Hard Fescue*) 46

FESTUCA RUBRA (*Red Fescue*) 48

LOLIUM PERENNE (*Perennial Rye Grass*) 50

LOLIUM VULGARE. L. ANNUUM (*Common*, or *Annual Rye Grass*) 55

LOLIUM ITALICUM (*Italian Rye Grass*) 56

PHLEUM PRATENSE (*Timothy*, or *Meadow Catstail*) 60

POA PRATENSIS (*Smooth-stalked Meadow Grass*) 62

POA TRIVIALIS (*Rough-stalked Meadow Grass*, or *Orcheston Grass*) 65

POA NEMORALIS (*Wood Meadow Grass*) 66

POA SEROTINA (*Late Meadow Grass*) 67

POA AQUATICA (*Water-Meadow*, or *Sweet Reed Grass*) 67

AGRICULTURAL CLOVERS (*Leguminosæ*)

PAGE

Trifolium repens perenne (*Perennial White Clover*) 68

Trifolium pratense (*Red*, or *Broad Clover*) 70

Trifolium pratense perenne (*Perennial Red Clover*, or *Cow Grass*) . . 71

Trifolium hybridum (*Alsike Clover*) 74

Trifolium minus (*Yellow Suckling*) 75

Medicago lupulina (*Common Yellow Clover*, *Trefoil*, *Nonsuch*, *Black Grass*, or '*Hop*'). 76

Lotus corniculatus (*Birdsfoot Trefoil*) 77

SUNDRY PLANTS USED FOR GRAZING, FEEDING, OR MAKING INTO HAY

Achillea Millefolium (*Yarrow*, or *Milfoil*) 78

Cichorium Intybus (*Chicory*, or *Succory*) 79

Medicago sativa (*Lucerne*) 79

Onobrychis sativa (*Sainfoin*) 82

Petroselinum sativum, or Carum Petroselinum (*Sheep's Parsley*) . . . 83

Poterium Sanguisorba (*Burnet*) 83

SOWING GRASS SEEDS 85

Spring Sowing 87

Sowing with Rape 90

Autumn Sowing 90

THE IMMEDIATE AFTER-MANAGEMENT OF NEW PASTURES 93

THE MANAGEMENT AND IMPROVEMENT OF OLD GRASS LAND . . 100

BREAKING UP OLD GRASS LAND 107

TEMPORARY PASTURES 110

One Year's Ley 114

Two Years' Ley 114

Three Years' Ley 115

Four, Six, or Eight Years' Temporary Pasture 115

HINTS ON HAYMAKING 119

HINTS ON GRAZING. 128

ENSILAGE (GRASS) 132

NOTES ON CERTAIN WORTHLESS GRASSES AND INJURIOUS AND POISONOUS PLANTS FOUND IN PASTURES 134

PERMANENT AND TEMPORARY PASTURES

From the Returns for 1909, issued by the Board of Agriculture, it will be seen that in the United Kingdom the total area of land under all kinds of crops, bare fallow, and grass, excluding mountain and heath land, is 46,885,810 acres. Of this area, 27,428,244 acres are in permanent grass, and 19,457,566 acres are arable; so that the land under permanent grass exceeds the arable by 7,970,678 acres.

Clover, Sainfoin, and Grasses returned under rotation crops augment the total of grass by 6,587,772 acres.

These facts demonstrate the immense national importance of this branch of agriculture.

The object of this volume is to offer suggestions for the improvement, where necessary, of meadows and pastures, and for maintaining both in a highly profitable condition. It is also hoped that those who contemplate laying down land to grass will find in the following pages information which may enable them to avoid waste of money, and at the same time ensure an adequate return for the necessary outlay, which can never be small where first-class results are aimed at.

One of the causes of the extension of grass since 1870 has been the grave difficulty concerning labour, and this difficulty becomes more acute in every succeeding year. The majority of able-bodied labourers and lads refuse to work in the country. Farmers are therefore increasingly dependent on the mechanic and the engineer. It has become necessary to take advantage of every invention and scientific discovery which reduces manual labour. Enormous strides have already been made in this direction. Threshing-machines, steam-ploughs, self-binders, spraying-machines, cream-separators, and facilities for transit have revolutionised farming conditions. Without the aid of machinery it must be patent that British husbandry could not be carried on by the present body of farm labourers. Finality in labour-saving has no more been attained in husbandry than it has in manufacturing industries.

Laying down land to grass has had a full share in solving the labour question, and has been largely instrumental in improving the economic position of agriculture. It has also been the means of preventing deterioration of land and of attracting tenants to farms. Yet it must be admitted that the extension of permanent pastures cannot be a national advantage, because grass land produces less food than does arable. And many large tracts of land are entirely unsuited for the formation of meadows or pastures; the finer grasses speedily disappear, while the soil is gradually filled with moss, weeds, twitch, and worthless indigenous grasses. But there is scarcely any farm land that will not profitably respond to the alternate system referred to in the chapter on 'Temporary Pastures,' and the advantages of this system are by no means restricted to soils which are unsuited for Permanent Pastures. The practice of the Lancashire and Scotch farmers has abundantly proved that no other method of farming pays so well as laying down the best land in artificial grasses for periods varying from two to four years, instead of simply sowing clover and breaking up after the first season. Temporary pastures are now recommended for general adoption by some of the most enlightened and able agriculturists of the country. A wide extension of this system is not merely a great means of lessening the labour bill, but it also ensures the storage in the soil of a large reserve of grain-producing energy ready for any national emergency.

The Drainage of Grass Land

It is beyond the scope of this work to treat the subject of drainage exhaustively. But in offering a few practical hints on draining operations, I propose to refer to some of the laws which render it impossible to maintain a first-class pasture unless land is properly drained either by natural or by artificial means.

An impression widely prevails that, however necessary effectual drainage may be for other farm crops, grass land may with impunity be left undrained. This is only one of the careless traditions which former generations have handed down concerning pastures. No one who is accustomed to examine water-logged meadows can fail to be impressed by the worthless character of the herbage they produce. If there were no cure, or the expense of the remedy were out of proportion to the benefit to be derived from it, apathy might be excusable. As to the greater part of the undrained grass land which is now lying in a comparatively unprofitable condition through stagnant water, there is no doubt that the vegetation can be immensely improved in quality. Drainage of pastures has never been known to be other than beneficial; and in most instances the quantity of hay or feed will also be greatly augmented, although from some land already yielding a large bulk of herbage of a low quality the increase after draining may not be immediately apparent. The total weight may even be temporarily diminished, but as the loss will be limited to those plants which possess little or no feeding value, regret need not be wasted on their disappearance. As a rule, the work can be carried out at a cost which will be returned with interest in the course of a very few years. Drainage alone will go a long way towards turning a marsh into a profitable pasture, and it renders other improvements possible at a trifling expense. The important point to be urged here is that undrained land should not be laid down to grass. Otherwise careful tillage, costly manures, and the finest grass seeds will certainly be wasted. The result is only a question of time. Sooner or later the valuable grasses which are sown will be supplanted by sedge and rush and other semi-aquatic vegetation, until the pasture gradually reaches the worthless condition which invariably prevails on undrained land.

Every year more water passes through land which is naturally or artificially drained than through soil which is generally saturated with moisture. Where stagnant water lies no rain can enter: it simply runs off the surface by any outlet it can find. The soil can neither breathe nor digest any fertiliser applied to it, and it is incapable of utilising the sun's heat for the development of plant life.

When rain falls on a well-drained field it does more than merely moisten the soil and supply plants with water. It has been computed that in each year, by means of rain alone, every acre of well-drained land in this country is benefited to the extent of five to ten pounds of nitrogen. One of the advantages of good drainage is that it allows the atmosphere to be freely carried into the soil by rain, when the oxygen sweetens and converts injurious organic substances into wholesome food for plants. At the same time, carbonic acid gas derived from rain and air performs the same operation for the mineral constituents of the soil.

Another advantage which results from draining is an increase in the temperature of the soil. It is well understood that evaporation produces cold, and the more rapid the evaporation the greater the cold. Travellers in the East will recall the delightful surprise experienced when first they drank cool water from a porous jar while the thermometer registered over 100°. Here is an illustration of the conditions which prevail on a hot day with water-logged soil. Under scorching sunshine, the soil when full of water becomes intensely cold immediately beneath the surface. The top crust may feel warm to the touch, but a plunging thermometer forced into the subsoil will reveal a difference of many degrees in temperature, which the rainfall on the surface is powerless to increase. When the sun's rays cease to fall on undrained land the cold subsoil quickly brings the surface to its own low temperature. This rapid change gives birth to the mists which in autumn are so familiar in the Fens and in the valley of the Thames. Surely there need be no wonder that under these adverse conditions the grass on badly-drained land is late to begin growing in spring and early to cease in autumn.

An eminent German authority has demonstrated that there is an intimate connection between a warm dry soil and economy in feeding cattle. Friable land absorbs more heat than land which is saturated with moisture, and retains the heat for a longer period. Upon the one animals lie warmer, especially at night, than they do upon the other. A large proportion of the food consumed by animals is utilised for the production of the heat which is constantly dissipated from their bodies. It follows that additional food becomes necessary to replace the animal heat lost by the colder surroundings.

Land which is properly drained comes under the influence of another operation of nature, to the great advantage of the crops upon it. Water would, after it has passed through the surface to the subsoil, be lost to plant life, were it not for the wonderful natural arrangement known as capillarity.[1] As the surface soil loses water by evaporation, it draws up and reabsorbs moisture from below; and this is especially the case when the soil becomes dry, and its particles are disintegrated. The water which is thus brought from the subsoil contains in solution some of the mineral constituents from the formation below, which further aid the growth of plants. This fact accounts for the widely different grasses which are to be seen in old pastures on surface soils that appear to be identical. The mineral constituents in a state of solution are brought up by the water from considerable depths, and by this means, amongst others, the geological substratum asserts its influence upon the herbage growing on the surface.

It is a mistake to suppose that the rainfall goes direct to the drains and is at once expelled from the land. On the contrary, the rain sinks into the land until it meets and mingles with the subsoil water, and the drains do not begin to run until the subsoil water rises above their level. While water, however small the quantity, is flowing in a drain-pipe—and probably long after it has ceased to flow—it may be taken for granted that the subsoil is saturated with moisture up to the level of the drains. The rise and fall of the subsoil water are therefore determined by the level of the drain rather than by the surface of the soil, as it would be in an undrained state. Thus, in well-drained land, the air is being continually carried into the soil by rain, and forced into it by atmospheric pressure as the subsoil water falls to a lower level, whilst the air is expelled

[1] Baron Liebig, in his *Natural Laws of Husbandry*, thus describes the action of water in a state of motion:—

'If we regard the porous earth as a system of capillary tubes, the condition which must render them best suited for the growth of plants is unquestionably this: that the narrow capillary spaces should be filled with water, the wide spaces with air, and that all of them should be accessible to the atmosphere. In a moist soil of the kind affording free access to atmospheric air, the absorbent root-fibres are in most intimate contact with the earthy particles; the outer surface of the root-fibres here may be supposed to form the one, the porous earthy particles the other wall of a capillary vessel, the connection between them being effected by an extremely fine layer of water.'

when the water rises. A water-logged surface is injurious to plant life not only because there is too much moisture in it and too little warmth, but because neither rain nor air can enter from above, nor mineral constituents be drawn from below. Drainage sets all these natural forces in motion, and they open the soil and disintegrate its particles for the benefit of plant life.

Again, drainage is always beneficial in promoting the early and late growth of grass: this alone is of enormous value in feeding stock. The early autumn and late spring frosts do not arrest growth on drained land so quickly as on that which is sodden with moisture. And on the latter there is also the additional injury which the hoofs of cattle inflict on the grasses. Thus one of the effects of drainage is to produce an ever-growing crop.

It has been urged with perfect truth that from arable land manures are often washed into drains, especially in wet seasons, and that in draining a farmer may be providing an outlet for manure which he has placed on the surface at great expense. Experiments by the late Dr. Voelcker and others have clearly proved that, with one exception, for which the remedy is easily applied, the loss of fertilisers by means of the drains is inappreciable when a green crop is on the ground; while water flowing from the drains under a bare fallow alongside may at the same time be highly charged with manurial matter. Hence the grass farmer is protected, as the arable farmer cannot always be, from this particular loss. The exception alluded to above is the possible loss of lime—an essential constituent of plants, and one of the substances most easily lost by the drains. This accounts for the necessity of applying lime from time to time on drained land which happens to be deficient in it. But while ammonia—which, by the process known as nitrification, becomes oxidised into nitric acid, and, entering into combination with lime, forms nitrate of lime—may possibly be wasted, it is satisfactory to remember that superphosphate of lime and other forms of phosphoric acid are never thus lost. Nor does it appear that potash is easily abstracted, so that there need be no hesitation in applying these substances from fear that they will be carried away by the drains.

It may be accepted as a general truth that grass land should not be drained so deeply as arable land. The weight of engines and heavy agricultural machinery has not to be allowed for, and there is no doubt that grass can advantageously absorb more moisture than corn crops. Further, the roots of most grasses do not penetrate very deep, and therefore it is desirable to have the water somewhat nearer the surface than on the arable part of the farm.

As to the practical part of draining I need say but little. There are tracts of country without any arterial drainage, no river or stream being available into which drains, if laid, could discharge their effluent. The remedy is generally beyond the power of private individuals. Work of this character can only be carried out by Government, or under parliamentary sanction by companies possessing large funds. This subject will, no doubt, one day claim attention from statesmen. Scientific engineering has rendered the task possible; the real difficulty lies in its cost. Meanwhile, in such exceptional districts, pastures having no fall for drain-pipes may be considerably improved by a system of gutter-cutting on the surface—a practice which would often be serviceable on land subject to floods. Water should not be allowed to lie on portions of a field after the main stream has retired sufficiently to enable this water to flow if only a channel were provided.

The manner in which drainage should be carried out in any particular case depends on soil, climate, and other circumstances, which must of course be taken into account, although they concern the details and not the principle of the work. The difference between the rainfall in the eastern and western counties,[1] or between the West of England and Ireland, will regulate the nearness of the lines of drains and the size of the pipes. But these differences do not touch the main question, whether to drain or not to drain. Soils which rest upon a porous subsoil certainly do not need draining. Other land may be retentive, and yet lie so high, or at such a steep inclination, that the water is discharged with sufficient rapidity without artificial aid. Indeed, draining may always be considered unnecessary where the best natural grasses grow luxuriantly. With these exceptions all clay lands, whether the clay is only in the subsoil or rises to the surface, and all peat lands, whether the peat has clay beneath it or not, and in fact all land which is habitually saturated with water, must be effectually drained before a pasture of any value can be established.

The prejudice occasionally expressed against the adoption of a system of drainage can generally be traced to some instance where the workmanship has been bad, or where no care has afterwards been taken to maintain the efficiency of the pipes, which ought not to be covered in until they have been proved to work satisfactorily. As draining is usually put out to contract, this matter needs close personal supervision.

It is a safe general rule not to make any single drain too long. Plenty of fall should be given, or the pipes may not

[1] The following statistics, compiled from the Reports of the Meteorological Office, show the difference in rainfall which occurred during 1909 in various districts in Great Britain:—

England (Highest rainfall)—Counties of Cumberland, Westmorland, Lancaster, Chester, Derby, and Stafford	38·87 inches
„ (Lowest rainfall)—Counties of London, Middlesex and adjacent on the North and East	24·95 „
Scotland—North-East	31·96 „
„ North-West	48·78 „
Wales	37·82 „

The case named by Professor Ansted in his *Physical Geography* is still more remarkable. He says: 'At Seathwaite the fall is 127 in., and a few miles off, at Bishop's Wearmouth in Durham, on the other side of the moors, it is only 17 in.'

work well after they have been laid some time. A good fall renders them to a considerable extent self-cleansing ; and the small drains should not enter the large drains at right angles, but always obliquely, so that the water may retain its momentum, when the flushing after a sharp storm will prevent the pipes from becoming choked. For short distances near hedgerows or trees, the use of socket pipes securely jointed with cement—not clay—is to be strongly recommended. The slight additional expense may save a large subsequent outlay.

As to the depth at which the pipes should be inserted, and the distance between the rows, no definite rule can be laid down. Experience has proved that in heavy land they must be near together, and not too deep ; but in lighter land the lines may be comparatively far apart. It is a common practice to cut the trenches three feet deep and to allow a distance of fifteen feet between the rows, but almost every field has some peculiarity of conformation or subsoil which affects the question. After the pipes are covered in, one man should be held responsible for periodical examination of the outlets, to ensure their being kept in working order.

An indurated pan, or hard mass, may have been formed beneath the cultivated surface by the weight of the plough and the trampling of horses during a long series of years. A similar condition, resulting from natural causes, is found on some heath lands, four or five inches below the surface. These hard subsoils are as impervious to water as beds of cement. Before putting drain-tiles into such land, an experiment should be made to ascertain whether satisfactory drainage cannot be obtained by breaking up the subsoil to a sufficient depth. If the trial prove successful, the expense of putting in tiles may be saved, and the fertility of the land will be increased. In many cases it may be necessary both to break up the subsoil and to lay in pipes before effectual drainage can be secured.

Cultural Preparations

Among the questions which need consideration before laying down land to grass, probably not one has received less attention than the condition of the soil. Grass is frequently regarded as a last resource for thoroughly exhausted land which no longer pays for the cultivation of any other crop. It is too often assumed that grass will grow anywhere, and under all circumstances. Although the old and wasteful process of allowing land to 'go to grass' is no longer advocated to any extent, many farmers consider it sufficient to harrow in a few seeds, and let them take their chance. The practice is extremely mischievous, resulting in immense loss of both time and money. No farm crop requires more care in the preparation of the land than permanent grasses; and there cannot be greater folly than to sow costly seeds, especially of the finer varieties, on land which has not been adequately prepared to receive them.

The choice of land suitable for permanent pasture is seldom open. Other circumstances than fitness for this purpose generally determine the matter. But occasionally it does happen that on some estate or farm there is a possibility of selecting the fields which are to be turned into grass. The guiding principles are few and simple. It may be accepted as an established conclusion, that sharp sands and gravels are not well adapted to the formation of pastures, but that heavy loams and most strong clays are eminently suitable for grasses and clovers, and will produce abundant crops. The fact that heavy soils are expensive to cultivate as arable is an additional reason why they should be laid down to grass. Again, if there be the choice of two fields, one sloping to the north and the other to the south, preference should be given to the former, because it will be less liable to burn in a hot summer.

Drainage has been referred to in the preceding chapter, and is a matter of the utmost consequence. If the land is naturally well drained, there will be a fortunate saving of expense, but otherwise this operation should be preliminary to all else.

Beyond question, the very best preparation for a spring sowing of permanent grass seeds is a bare fallow in the previous summer. This affords the opportunity of destroying successive crops of indigenous annual weeds, and it is important that these should be got rid of by scarifying and dragging rather than by ploughing, for the plough is only too certain to bring to the surface a fresh stock of weed seeds ready to germinate in the following spring. Many influences may aid or hinder the work, which depends not only on the character of the soil and the previous cropping, but also on the atmospheric conditions which prevail while the operations are in progress. Here the advantage of a bare fallow is realised, because the cultivator has the whole summer and early autumn in which to accomplish the task.

Deep ploughing should be carried out first, and if subsoiling is considered necessary, there is all the greater reason for doing it early. Then, by means of the scarifier and the roller, the soil can be cleaned and so far rendered fit to receive the seeds that in the following spring only one or two turns with the harrow will be required to perfect the seed-bed. There are good reasons for insisting on thorough preparation of the land in the first instance. Careless and half-hearted work wastes both seed and labour, and the routine has to be attempted a second time under great disadvantages. Causes entirely beyond human control may sometimes render it needful to re-sow, even after the most earnest effort; but no one should lay himself open to the possibility of self-reproach for having contributed to partial failure by neglect. More of the failures in attempting to create pastures could, if all the facts were known, be traced directly to the unfavourable state of the soil, and to its previous cultivation, than is generally believed, and it is true wisdom, as well as sound economy, to wait a year, or even two years, rather than risk sowing upon soil which is foul or out of condition.

The bare fallow, however, will be the exception; as a rule it cannot be afforded. In the interests of the coming pasture, a root or potato crop is the next best preparation, and unless the land is capable of growing a first-class crop of roots, it will be incompetent to produce even a fair pasture. A root crop offers the advantage that while few are disposed to manure a bare fallow heavily, a liberal dressing of farmyard dung will not be denied to the mangels or swedes. To young grasses it is a great gain when the land can be made rich and put into good heart in the preceding year, instead of manuring immediately in advance of sowing seed. The tender and delicate roots of the rising grass plant may be seriously impaired by contact with crude raw manure; whereas they will readily assimilate a rich dressing which has had time to become mellow, or to be absorbed into the staple.

Supposing land to be prepared by feeding off a crop of turnips with sheep, it may happen that the turnips have to be supplemented with meadow hay. If so, it is important that the hay should be only such as has been cut very early, otherwise the ripe seeds of the least useful grasses will, as they pass the sheep undigested, in due time spring up and make the pasture foul. Such grasses as *Holcus lanatus*, *Bromus mollis*, and other worthless varieties, often find their way into a pasture in this manner.

Whether the roots are fed off during September or October by sheep eating hay or cake—and the use of cake is to be strongly commended—or whether the roots are carted off during autumn, in either case the plough should be put into the ground the moment it is at liberty. This first ploughing must be deep and thorough, and should be quickly followed by another ploughing to lay the land up rough for the winter. In February, or as early as the land is workable, get the harrow and the roller upon it until the seed-bed is fine, firm, and level. A tenacious soil, which dries off lumpy, may involve the expenditure of much time and energy to put it into good order. The delay will prove tantalising, but impatience is a bad husbandman, and the implements must be kept going until a satisfactory finish is obtained. It should be remembered that few grass seeds will grow at a greater depth than half an inch, even in fine friable soil. In cracks and fissures they will be utterly lost. Hence a sowing on ground which is rough is foredoomed to partial or entire failure, and the plants which do come will be the coarser varieties only.

Consolidation is equally important, for the young grasses cannot obtain foothold upon a loose or hollow soil. In such a case it is impossible to secure a perfect plant; and here again the finer sorts will fail. It is no unusual thing to see a full plant of grass all round the headlands of a newly-sown field, while the centre is thin or bare. The explanation is that the greater traffic over the headlands created a firmer seed-bed for the grasses than was made for them elsewhere.

Even after land has been fully prepared for the seeds, it will be all the better if allowed to lie untouched for a few days before sowing; but if the season is advanced waiting may be dangerous. Otherwise the delay offers two advantages. It allows the soil further time to settle down, and also gives the annual weeds a chance to start, so that by a final turn of the harrow they may be killed before the grass seeds are sown.

Annual weeds, unfortunately, are certain to come only too plentifully, and will demand constant attention when grass seeds are sown without a corn crop in spring.[1]

As a preparation for autumn sowing, no other crop is equal to an early variety of Potato. Earthing-up the rows exposes a great surface to atmospheric influences, and this materially aids the disintegration of the soil. Another point in its favour is that the crop is generally lifted by hand, and the soil is subjected to a course of spade husbandry, which, as a preparation for grass, is superior to all other modes of cultivation. When digging the Potato crop the labourers should fork up and throw aside every bit of couch they come across. This will very effectually assist the cleaning process. The only objection

[1] A remarkable instance of the vitality of buried seeds was reported by a correspondent in *The Times* of March 26, 1894. The following is a verbatim copy of the letter:—

'The fact that seeds of weeds, especially of annuals, are capable of retaining their vitality for a number of years in the soil is a circumstance well known to farmers of arable land. An objection often advanced against deep ploughing, particularly of light soils, is that it may "bring up the charlock." To what length of time buried seeds may continue alive is not certainly known, but the following carefully made observations, sent to us by Mr. S. James A. Salter, F.R.S., of Basingfield, near Basingstoke, are well worth recording:—"Twenty-four years ago I purchased this property (Basingfield), a large portion of which was at that time arable land bearing good crops of grain, which were, however, in a very weedy condition, the principal weeds being charlock (*Sinapis arvensis*), red poppy (*Papaver Rhœas*), and fumitory (*Fumaria officinalis*). It is important to note that all three of these plants are annuals. In the autumn of that year I laid the land down to grass (permanent pasture), and it has remained so ever since. I make hay annually with the first crop of grass; and the second I feed off with sheep. From time to time I dress the grass with artificial manure—kainit and superphosphate. None of the before-named weeds are ever seen unless the soil is disturbed; but directly the surface is broken, and soil some six or eight inches deep is brought up and exposed to atmospheric influences and light, all three of the weeds named appear in abundance, especially the charlock and poppy. This occurred conspicuously last year. The seeds producing these plants had been buried twenty-three years, but at a depth beyond vivifying influences, though still retaining vitality. There can be no fallacy in the observation; it has occurred over and over again. Eight years ago, after a very hard frost and a thaw, the surface of the ground being very rotten, I had occasion to take a waggon heavily laden across this pasture; the wheels sank deeply into the soil, and tore up the ground, bringing to the surface much subsoil. In the spring these furrows were filled with charlocks, and presently, when they flowered, there were two parallel yellow ribands to be seen across the land, following the irregular course the waggon had taken. It was a most striking sight. There was not another charlock to be seen in the field. The seeds producing these beautiful yellow ribands had been buried fifteen years."'

to sowing immediately after Potatoes is the difficulty of consolidating the land; but by planting a first-early variety the crop can be marketed in July, and before the grass seeds are sown in August a persistent use of the harrow and roller will do much to make a firm seed-bed.

It will now be proper to refer to some of the emergencies which arise when land must be laid away to grass at the earliest possible moment, whether in a fit condition for the purpose or not. One of the commonest instances is that of a clover ley which it is desired to turn into a permanent pasture. There is a natural feeling of reluctance to break up the clover plant, and the hope is indulged that grass seeds will take upon it. The objections to this course are many and serious, although they are not always insurmountable. Possibly indigenous weeds have already such a hold of the ground as to afford very little chance of the grasses making head against them; and in soil crowded with clover roots the young grasses have not a fair opportunity of establishing themselves. Still, however undesirable the practice of turning a ley into a permanent pasture must ever remain, necessity knows no law, and sometimes this unpromising experiment is crowned with success. Those who leave much to chance will deserve and obtain a poor result; but the man who is persistent and determined to succeed will often secure an adequate return for his labour and outlay. The chief inducement to make the attempt is the probable saving of a considerable expense in breaking up the land and getting it ready to sow down again.

Vigorous harrowing in the autumn is the first process in converting an old ley, and it must be no child's play. There is not the least cause for alarm in the apparent wreck of the standing plant. The more ruthlessly it is torn the better chance will there be for the grass seeds, and the more satisfactory the ultimate pasture. Follow up with a top-dressing of cake-fed manure or compost early in the winter, and the land will then, although only in a limited and imperfect degree, be prepared to receive the grass seeds in the following spring. Of course the old clover plants afford shelter to the newly-sown grasses.

So little remains to be said on this part of the subject that perhaps it will be convenient to dispose of it at once, although the question of seeding does not properly belong to this chapter. As to the choice of seeds, it is mere waste to sow fine or weak growing varieties on an old clover ley. The adverse circumstances of the case will afford them little chance of struggling into life, to say nothing of a profitable existence. The sorts selected must be the stronger and more robust of the perennial grasses, and the seed should be got in early, before the clover has time to shoot vigorously in spring. Accomplish the task in February if possible; bush-harrow after sowing, and as a finish put the roller twice over every part of the field.

Another plan I have personally adopted, with success, has been to feed the second crop of a clover ley in August with sheep eating cake, sowing the grass seeds among the clover in front of the sheep, and leaving them to trample in the seeds and manure the land as they eat off the crop. It is important not to sow in the early morning, or at any time when the standing plant is wet, or the seeds will adhere to the clover and be eaten by the sheep, instead of falling to the ground to be trodden in.

Other instances of a similar character might be cited, but as they only need some modification of the method already explained, it may be enough to say that I have known tolerably successful pastures to be formed on an old Sainfoin ley, a worn-out Lucerne plant, a three or four years' ley, and even on clean Barley and Oat stubbles, without ploughing or using any other implement than the harrow, the seed-barrow, and the roller.

A different but very frequent case of emergency is that of a piece of glebe or other land that has been neglected for many years until it has become a perfect mat of black twitch (*Alopecurus agrestis*). In despair of cleaning it at a reasonable cost, the rector or owner decides to allow it to 'go to grass,' as hundreds of acres have actually gone. Wisely, it is considered desirable to give Nature some assistance, but it is almost a misnomer to dignify that assistance by the name of preparation. In this instance also the routine previously advised is applicable. Rigorous harrowing in autumn, a heavy top-dressing during winter, and the sowing of suitable strong-growing seeds in early spring, are the means by which the most profitable results can be ensured. I know many cases where this rough-and-ready treatment has been followed by a fairly paying plant. Especially may improving crops be anticipated when the land is continuously manured, or where grazing cattle are liberally assisted with artificial food.

The Selection of Grasses and Clovers

All the operations which concern the making of a pasture are important, but it is no exaggeration to say that a judicious selection of the various grasses and clovers which are to constitute the crop may be justly regarded as vital to success. Failure here means the waste of all other energies, for it is worse than useless to incur the labour and expense of establishing plants which are not wanted. However good they may be elsewhere, they will be no better than weeds if they fail to answer the required purpose. The choice of suitable seeds has provoked greater conflict of opinion, both among theorists and practical men, than aught else, and in my opinion the main cause of the controversy arises from the attempt to deduce large inferences from small experience. The laying down of land to grass is only an occasional incident on most farms—perhaps it would be correct to say on most estates. It is the exception to find persons who are able to speak from experience based upon actual practice over more than a very limited area. Yet the man who has achieved a single success in laying down land may write to a daily or weekly newspaper, or deliver a speech at a local farmers' club, from which it might be assumed that the agriculturists of the United Kingdom will find in a particular mixture of seeds the preventive of all the ills to which grass lands are subject. A little knowledge on laying down land is a very dangerous thing. No prescription, however excellent every one of the varieties which compose it may be, can by any possibility be suitable for universal application. The attempt to put forward even a first-class mixture of grasses for all soils and all purposes savours essentially of empiricism. Those who possess the widest experience on this subject are least inclined to lay down rigid rules. Land agents who have had the management of large estates in various parts of the country, and who have had greater opportunities for extensive observation than most men, are exceedingly careful to consider differences of soil, subsoil, and the purpose to which each individual pasture is to be devoted; and their success is chiefly attributable to the wise application of general knowledge to special cases.

It may be interesting to recall some of the extraordinary statements which have been made in public journals and at local farmers' clubs as to the value of certain grasses. Several years ago it was confidently declared that Italian Rye Grass should not only be grown alone and in alternate leys, but that no permanent pasture could be successful which did not contain a large proportion of it. After this assertion had been fully discussed and proved to be erroneous, there was a rebound to the other extreme. It was freely affirmed that Italian Rye Grass degenerated into twitch, and was therefore unfit for cultivation. There can scarcely be a doubt that samples of this grass containing seeds of twitch have been sown, but this pest has never been produced by pure seed of Italian Rye Grass. I entirely agree with those who hold that Italian should as a rule be excluded from mixtures which are intended to form permanent meadows or pastures, and I also concur in the general consensus of opinion among practical men that as a forage plant Italian Rye Grass has a high value either sown alone or as a component of ley mixtures.

Among the grasses which have provoked great diversity of opinion, Cocksfoot affords a conspicuous example. Instead of being reserved for those soils and purposes for which it possesses an undoubted value, it has been recommended for the production of high-class hay everywhere, and even for sowing on geological formations for which it is totally unsuited. Excellent pastures have been literally ruined by the introduction of Cocksfoot. Alternate leys on the Chiltern Hills, where Rye Grass and Clover had previously answered well, have, by the introduction of Cocksfoot, yielded almost unsaleable hay, and, having once been allowed to seed, the Cocksfoot has proved difficult to eradicate. Those who are familiar with the art of forestry are well aware that it would be futile to attempt to grow elm timber on sandy land, or larch on land which is only fit for Scotch fir. Had the laying down of grass received the careful attention which has been devoted to forestry, it would be considered just as unreasonable to sow any one species of grass on all soils and for all purposes, as to recommend elm trees to be planted everywhere.

The sowing of Poas has been condemned as unnecessary because it happened that some varieties of Poa were indigenous where the experiment was made. But a New England farmer will not hesitate to sow *Poa pratensis* alone, and long experience has proved that he does not prize this grass too highly. On the other hand, in certain districts of New Zealand, where *Poa pratensis* develops a troublesome twitch-like habit of the worst character, the suggestion that it should be sown at all would only provoke an expression of incredulous amazement. Surely such a widely different estimate of the value of a single variety may well suggest a doubt as to the universal adaptation of any one kind of grass to all soils and districts. Those who possess the largest knowledge, obtained from the widest sources, agree that each individual case must be considered independently and on its own merits. From the beginning there should be a clear understanding of the condition and capabilities of the soil. The subsoil, too, must be taken into account, for sooner or later its influence will tell decisively upon the existence of certain grasses. Then the purpose of the crop must not be overlooked. Whether it is chiefly for hay, or entirely for grazing, will prove an important consideration in determining the sorts to be sown. Even the kind of cattle the land is intended to carry is worth more than a passing thought. Milch cows, fatting stock, sheep and horses, or a combination of these animals, can be provided for, if proper care be taken in the selection of grasses and clovers.

In succeeding chapters the various plants, seeds of which are generally sown for permanent pastures and alternate husbandry, are described, with their relative value, fitness for certain purposes, and such other particulars as indicate the sorts and proportions to be used on the different soils and under the conditions usually prevailing in this country. In so complex a subject it is not easy to make the selection in any case. But I am anxious, as far as may lie in my power, to prevent a repetition of the costly blunders which have too often been made in this branch of British agriculture. With this object in view six typical prescriptions of Grasses, Clovers. &c. for various soils are given on the following page.

TYPICAL PRESCRIPTIONS OF GRASSES AND CLOVERS.

In the following typical prescriptions of Grasses and Clovers the figures refer to the proportions of seed by number, not by weight. This is the only method of stating accurately the proportions of plants which are suitable for any kind of soil. No two varieties contain a similar number of seeds per pound, and in some cases the difference is very wide indeed. This will be instantly understood by comparing samples of two Grasses such as Meadow Fescue and Timothy, or Cocksfoot and Smooth-stalked Meadow Grass. The same difference is found in Clovers, as may be seen by comparing White Clover with Cow Grass. As to the weight, the tables below assume a total seeding of 40 lbs. per acre, and that each variety is the finest seed produced by the last harvest, perfectly pure and proved to be of the highest germinating power.

Varieties of Grasses and Clovers		Rich Deep Loams	Medium Loamy Soils	Light Loams	Heavy Clay Soils	Brashy Soils over Chalk	Sheep Downs
		Per cent. per acre	Per cent. per acre	Per cent. per acre	Per cent. per acre	Per cent. per acre	Per cent. per acre
Achillea Millefolium	(Yarrow, or Milfoil)	—	0·9	1·0	—	0·9	0·9
Agrostis stolonifera	(Fiorin)	1·5	1·0	1·1	1·5	—	—
Alopecurus pratensis	(Meadow Foxtail)	11·2	9·3	8·0	13·6	7·7	—
Anthoxanthum odoratum	(Sweet Vernal)	1·1	1·1	1·9	—	—	1·0
Avena flavescens	(Yellow Oat Grass)	2·5	2·3	2·7	—	2·5	4·5
Cynosurus cristatus	(Crested Dogstail)	1·4	1·3	2·6	1·4	2·8	3·8
Dactylis glomerata	(Rough Cocksfoot)	13·0	9·3	6·1	14·5	5·8	1·9
Festuca duriuscula	(Hard Fescue)	5·0	5·3	6·4	3·5	7·8	13·1
Festuca elatior	(Tall Fescue)	4·2	2·2	—	4·8	1·2	—
Festuca ovina	(Sheep's Fescue)	2·5	2·3	5·9	2·5	7·5	11·3
Festuca pratensis	(Meadow Fescue)	8·7	8·6	8·2	6·9	7·5	5·7
Festuca rubra	(Red Fescue)	—	5·1	6·6	—	6·4	7·7
Lolium perenne	(Perennial Rye Grass)	12·0	11·5	12·6	13 5	14·6	12·5
Lotus corniculatus	(Birdsfoot Trefoil)	—	—	0·9	—	0·9	0·8
Lotus major	(Greater Birdsfoot Trefoil)	0·5	—	—	0·5	—	—
Medicago lupulina	(Yellow Trefoil)	2·4	1·0	3·1	2·0	2·4	2·9
Medicago sativa	(Lucerne)	—	—	—	—	1·0	—
Onobrychis sativa	(Sainfoin)	—	—	—	—	0·1	3·6
Carum Petroselinum	(Sheep's Parsley)	—	—	—	—	0·6	1·0
Phleum pratense	(Timothy)	11·6	10·8	5·5	11·7	5·8	2·7
Poa nemoralis	(Wood Meadow Grass)	—	2·9	2·9	—	3·1	2·8
Poa pratensis	(Smooth-stalked Meadow Grass)	—	5·5	11·3	—	9·9	10·9
Poa trivialis	(Rough-stalked Meadow Grass)	11·2	6·9	—	11·3	—	—
Trifolium hybridum	(Alsike Clover)	4·4	2·4	2·5	5·3	—	—
Trifolium minus	(Suckling Clover)	—	—	—	—	2·5	1·6
Trifolium pratense perenne	(Perennial Red Clover)	4·1	3·9	3·8	4·2	3·5	1·3
Trifolium repens perenne	(Perennial White Clover)	2·7	6·4	6·9	2·8	5·5	10·0
		100·0	100·0	100·0	100·0	100·0	100·0

Agricultural Grasses (Gramineæ)

AGROSTIS ALBA—VAR. STOLONIFERA

(*Fiorin*, or *Creeping Bent Grass*).

This plant thrives in spongy soil which is not firm enough to produce ordinary herbage, and in land which cannot be drained it will get a living where other grasses perish. In mountainous districts where rain falls frequently and abundantly, and the atmosphere is moist, Fiorin grows freely, both on light land and on peat. The plant affords very early feed in spring, but the chief value lies in its continued growth late in autumn. Fiorin has been pastured as late as the middle of December. If not eaten down in autumn the herbage in the following spring is sweet and wholesome food for young stock. One disadvantage is the liability of the panicles to become ergoted; another, that where Fiorin exists in large proportion the turf looks brown for some time after the hay has been cut.

Although this grass has the peculiarity of rooting from the procumbent nodes of the stem, especially in pastures much trodden by cattle, it is not dependent alone upon the surface soil for support. In suitable situations the roots penetrate the subsoil to a considerable depth, and a series of rainy summers always creates a demand for seed out of proportion to its value, but when hot dry years return Fiorin is quite as unreasonably condemned.

The Agrostis family is generally deficient in nutritive value, and although *A. alba stolonifera* is regarded as an exception, and has been highly recommended by several authorities, I do not concur in the opinion that it should invariably be included in ordinary prescriptions, if only for the reason that the true variety is not always to be depended on from a sowing of seed. The method of securing Fiorin by obtaining plants from land where it grows indigenously, cutting up the long trailing roots, and planting them in prepared ground, is far too costly a process.

Some of the other forms of Agrostis, such as *A. vulgaris*, *A. dispar*, *A. capillaris*, *A. alba*, and *A. nigra*, are known to be inferior to the true Fiorin, and the seeds of these cannot always be certainly distinguished, even by a botanist, from those of *A. stolonifera*. *A. vulgaris* is the Common Bent, and is usually regarded as a weed grass, although it is occasionally sown for a purpose outside farming. *A. alba* is the worthless Marsh Bent, and is never sown: neither is of the least value as fodder.

In the absence of the flowering panicle, *Agrostis alba stolonifera* is easily recognised at all seasons of the year by the short leafy branches which burst through the sheath at each node of the numerous stolons. The ligule is long and pointed, and sometimes serrated at the margin.

For illustration, description, and chemical analysis, see Eighth Edition.

ALOPECURUS PRATENSIS

(Meadow Foxtail).

Of the native species of *Alopecurus*, this is the only one which is used for agricultural purposes, and it is justly regarded as one of the most important grasses we possess. Cattle show great partiality for it, and chemical tests reveal its high nutritive qualities. Foxtail is thoroughly perennial in character, and does not attain complete development until three years after seed is sown. The full value of the plant cannot, therefore, be realised in alternate husbandry unless the ley is to remain down more than two years. Even then Cocksfoot is preferable for temporary pastures, although it is less nutritious than Foxtail. The latter is, in fact, essentially a landlord's grass, for which he may willingly pay in the full assurance that the future of the pasture will justify the outlay.

Foxtail begins to bloom about the middle of April, but by the third week of May it is practically in full flower, and should there be a sufficient proportion of it in a meadow to warrant early mowing, the crop ought to be cut; for although ripening does not, as is the case with nearly all other grasses, seriously deteriorate the quality, it is wasteful to allow the bulk of its palatable and nourishing herbage to be shrivelled up and lost on the ground while other varieties are maturing. The necessity for early cutting should influence the proportion of Foxtail included in a prescription for a meadow, but in laying down a pasture this consideration need not be entertained. After taking a crop, it is one of the quickest grasses to commence growing, shooting up its green herbage before other varieties make a start, and in bulk the aftermath frequently exceeds the early growth.

One characteristic which gives Meadow Foxtail a high value is the immense amount of leafy herbage produced in proportion to the quantity of stalk, and for this reason the extent of its presence in a pasture often remains unsuspected. The early growth is also in its favour. Other conditions being equal, a pasture which contains Foxtail in abundance will carry stock ten days in advance of a pasture in which it is wanting. These facts have doubtless suggested the idea that Foxtail is peculiarly a pasture grass, but it yields so good a crop, of such excellent quality, that it is almost of equal service for making into hay. As one of the few grasses that thrive under trees, it should be plentifully used in sowing down orchards and shady pastures.

In the North, Meadow Foxtail is not much sown for hay, because in high latitudes the culms seed long before other herbage is fit for the scythe. Otherwise this grass is as well adapted to Scotland as to the North of Europe generally. There is scarcely a forage plant known which endures cold so well, and spring frosts do it little harm. Even in the severest winters, when other grasses suffer much damage, Foxtail remains comparatively uninjured. The distribution of the plant in this country is very unequal. In Devonshire it is uncommon, and in South Wales rarely seen, but it is plentiful in most of the alluvial meadows bordering the greater rivers of England. Where this grass is entirely wanting I should scruple to recommend the sowing of any quantity in new pastures; but the great excellence of the herbage renders a trial on a limited scale desirable.

Sometimes Foxtail is carelessly mistaken for Timothy, but, besides other differences, the former comes to maturity quite a month before the latter. Both flourish in strong soils, and a stiff loam or clay is necessary to fully develop the fine qualities of Foxtail, and to maintain the herbage in a green state during hot dry weather. On well-drained retentive land the growth is very luxuriant, and during prolonged rain a heavy crop may be beaten down, when the plant is liable to rot at the roots. This fact again points to the necessity of mowing in good time. Although it is useless to sow seed on undrained land, this is one of the best grasses known for land under irrigation, and the water-meadows in the South of England, in which it grows freely, produce immense crops.

On very dry soils Foxtail is so stunted and diminutive as almost to lose its distinctive characteristics; but even under such conditions it may sometimes be worth using in small proportion, and wet seasons will prove that this is not a mistaken policy. In parks and paddocks surrounding residences, where a green appearance and constant growth are important, seed should be sown freely. The result will gratify the eye of the proprietor and benefit the cattle which graze upon the pasture.

True seed of *Alopecurus pratensis* is generally expensive, and it is so light and delicate in formation as to need exceedingly well-prepared land to ensure vegetation. Unfortunately, the stiff soils which specially suit Foxtail are the most difficult to make fine in time for sowing. But for this grass alone it is worth a strenuous effort to get the land into good order.

The short, degenerated, truncate ligule, and the long, broad, prominently veined leaves, together with the involute character of the emerging leaf, distinguish this species in the absence of its flowering spike-like panicle. A further aid to identity is the dark brown or violet sheath.

Experiments at Rothamsted and Kidmore prove that nitrate of soda and mineral salts have a marvellous effect upon the growth of Foxtail. It assimilates larger quantities of manure than many other grasses, and differs from Cocksfoot in the fact that the latter appears to benefit more from ammonia salts than from nitrate of soda, whereas Foxtail does equally well with both.

Most of the seed imported into England comes from Southern Russia. The German seed sold at Breslau is immature and grows badly. Occasionally very fine samples are obtained from Sweden, but the best seed of all is Dutch ; and although the most careful botanist can detect no difference between the seed grown in Holland and that obtained from other countries, the Dutch seed produces a much more robust as well as an earlier plant.

For illustration, description, and chemical analysis, see Eighth Edition.

AMMOPHILA ARUNDINACEA (Host.)
PSAMMA ARENARIA (R. & S.)
(*Sea Reed, Marrem,* or *Mat Grass*).

An Act passed in the reign of Elizabeth, and renewed in the reign of George II., prohibited the pulling up of this plant, or the destruction of it in any other way. Like

Elymus arenarius, it is used for fixing sand on the sea-coast, but it possesses no feeding value, and is rejected by all herbivorous animals. The name of Mat Grass arises from its being employed as a material for mats. It also makes excellent thatch. The time of flowering is July.

ANTHOXANTHUM ODORATUM

(*Sweet-scented Vernal*).

A very distinct species, grown chiefly for the pleasant odour it imparts to the hay crop, which enhances the price of the hay. It is only when the grass is dry that the fragrance is fully distinguishable. So marked is this quality that in the South of Europe an extract is obtained from the plant which is used in the manufacture of scent. This is one of the earliest grasses to start growing in spring, and it comes into full flower at the beginning of May. The plant is a true perennial, exceedingly hardy, and continues growing until late in autumn, so that the aftermath is actually much larger in bulk, as well as more nutritious, than the first crop. This fact clearly indicates that Sweet Vernal is quite as valuable for grazing as for cutting. Another characteristic of this grass is the brilliant green colour of the herbage; hence seed should be included in mixtures for ornamental grounds in larger proportion than would be reasonable if the quantity of hay or grass were of primary consideration; while, for the park, especially round a mountain home, it is invaluable. But to sow the seed for lawns is a mistake, although it has been recommended for that purpose. The leaves are too broad and flat, and look unsightly in a sward that is kept down close.

Sweet Vernal grows abundantly in some parts of Devon[1] and in the Eastern Counties, and the plants invariably thrive

[1] A gentleman in South Devon has written me concerning *Anthoxanthum odoratum* as follows:—

'This grass grows wild in this neighbourhood in the hedgerows, and cattle are better in pastures among other varieties than when two of them happen to be contiguous. On deep, rich, moist soils this grass grows luxuriantly, and on marshy land it throws up flower-culms all through the summer. In wet, peaty land the growth is so large and reed-like as to deceive those who are not experienced botanists. Sweet Vernal also thrives under the shade of trees and does remarkably well under irrigation. Yet, strange to say, it grows freely on thin moors and sandy dunes, and appears to be capable of living in soil which will not support any other useful grass.

When out of flower the plant is easily recognised by its sweet-smelling, short, broad leaves, as well as by the fringe of long hairs which surround the neat, obtuse ligule. The leaves, too, are somewhat hairy on both upper and lower surfaces. By masticating the stalk a lavender-like flavour can be detected.

The agreeable perfume imparted to hay by Sweet Vernal has created a demand for seed out of all proportion to the supply. Seed is gathered by hand from plants growing wild in the woods and clearings of Central Germany, and only a very small quantity is sent to this country. Hence the frequent substitution of the annual species *A. Puelii*, which is every year largely exported from Hamburg under the name of Sweet-scented Vernal. This grass (*A. Puelii*), which is not a native of Britain, cannot be made partially permanent even by the process of cutting, for it flowers all the summer long, and if prevented from seeding at hay time, the flower-heads are thrown up later, and, as cattle do not relish them, seed is produced and the plant dies. It is therefore extremely important to secure the true perennial *Anthoxanthum odoratum*, which is, and always must be, a costly seed.

fond of it in the winter months. It has the merit of shooting very early in the spring. Towards the end of February, after a few days of mild and moist weather, it commenced to sprout strongly, and I observed that all kinds of cattle preferred it to other grasses.'

Although manure is not directly injurious to Sweet Vernal, indirectly it destroys this grass by enabling other varieties to elbow the plant out of existence.

For illustration, description, and chemical analysis, see Eighth Edition.

AVENA FLAVESCENS

(Yellow Oat Grass).

This grass is not nearly so well known or appreciated as it deserves. It prospers in marl and calcareous soil, and in all light land rich in humus, particularly in that which contains lime. The forage is of good quality, and is greedily taken by cattle.

Avena flavescens does not flower until midsummer. In the Thames Valley it contributes no inconsiderable portion of the herbage of the water-meadows, and over the South of England it affords a fair quantity for the scythe at hay time, while the aftermath is abundant. The leaves are pale green; flowers golden yellow, and they show conspicuously.

In the absence of the flower the plentiful supply of soft deflexed hairs upon the lower sheaths, together with the ciliated ligule, are characteristics by which this grass may be recognised. In a favourable light the low acute ribs of the blade are clearly visible, and a single row of hairs is discernible along each rib.

The seed of Yellow Oat Grass is always expensive, and is sometimes difficult to obtain in sufficient quantity. Probably many of the adverse opinions expressed concerning this variety have resulted from sowing in its stead that pestilent weed *Aira flexuosa* (Wavy Hair Grass), the seed of which is not altogether dissimilar from that of *Avena flavescens*, but has often been sold for it by those who are not familiar with the true article. Seed of the latter, however carefully saved, only germinates moderately well, and the high price necessitates its sparing use. When arranging the prescription for a pasture a full seeding of other grasses must be given, and then, if expense be no object, a pound or two of *Avena flavescens* may be added.

Both at Rothamsted and Kidmore this grass has shown itself capable of holding its own very fairly without any manure at all. Nitrate of soda and mineral manures, however, considerably augment the produce.

For illustration, description, and chemical analysis, see Eighth Edition.

AVENA ELATIOR

(*Holcus avenaceus ; Arrhenatherum avenaceum*).

Tall Oat Grass, or False Oat Grass, is known by all these names; and in Dauphiné, where it is very much grown, especially for seed, it is called *Fromental.* Sinclair recommended this grass for agricultural purposes; but, notwithstanding his favourable opinion, it has not in this country received the attention it deserves. At one time, no doubt, its value was overestimated, and, in the reaction which naturally followed, the grass was consigned to undeserved neglect. Possibly the somewhat bitter taste, and the aversion which cattle at first display towards it when grown alone, may account for the indifference shown to it in England, although when mingled with other grasses the slightly bitter flavour is not apparent. On the Continent, and especially in Sweden, where it is largely cultivated, both cattle and sheep take it freely. Its value for cold northern countries is indisputable, but a warm climate is necessary to fully develop its great merits. In the Southern States of North America it is known by the names of Evergreen Grass and French Rye Grass, and is regarded as one of the very finest grasses for producing an immense weight of herbage, far surpassing Timothy in the abundance of its crop. The roots penetrate so deeply into the soil as to enable the plant to withstand both cold and drought.

Tall Oat Grass may be grown almost everywhere, but it is seen to least advantage on poor thin land. On good light and medium soils, as well as in all forms of clay, if not too damp, it grows from two to four feet high. In warm forcing situations it produces two heavy crops of hay in one season, and will continue to throw a blade until autumn frosts appear.

The plant starts into growth very early in spring, and after the crop has been mown there is one peculiarity which must be borne in mind, or much waste may result. When cut this grass absorbs moisture in the same manner as *Trifolium incarnatum*, and therefore hay containing it must be got into rick as rapidly as possible. As the stems are not succulent, *Avena elatior* can quickly be made into hay, which keeps particularly well.

For permanent pastures this grass cannot always be regarded as suitable, on account of its uncertain duration on some soils, and there is also the objection that the herbage is extremely coarse.

The long, smooth, narrow, deep green leaves, the youngest of which are tightly twisted or rolled throughout their length, together with the truncate hairy ligule and keeled sheath, enable a student to determine this species when the flowering panicle is absent.

The rapid growth of *Avena elatior* makes the plant a gross feeder, and it will absorb any reasonable quantity of manure, especially of the nitrogenous class.

When sown in autumn a much larger produce is obtained in the following year than from a spring sowing.

A weed variety of this grass, growing in arable land, is characterised by the formation of a bulb-like growth in the ground, just above the root, and is known by the name of Onion Couch. Tall Oat Grass has no resemblance to *Avena fatua*, the Wild Oat Grass or Havers, which is a weed of cornfields, and is much like the cultivated oat in appearance.

BROMUS INERMIS

(*Awnless Brome Grass,* or *Hungarian Forage Grass*).

In the South-east of Europe Hungarian Forage Grass is much used and produces an extraordinary amount of fodder. The plant is also largely grown in several parts of North America, where it endures both heat and cold, thriving where no other grass of any value can be relied on. In this country it is not indigenous, but all kinds of stock eat it greedily, even in preference to Italian Rye Grass. Compared with that grass, *Bromus inermis* starts earlier in spring, yields quite double the crop at the first cut, and the analysis made by Dr. J. Augustus Voelcker shows the *Bromus* to be the richer in albuminoids and nitrogen.

The plant is perennial, grows rapidly, and yields an immense quantity of succulent herbage. Seed is usually sown alone for a forage crop, but unfortunately the germination in the open is rather capricious, even after the usual tests have proved satisfactory.

BROMUS SCHRÆDERI

(*Schræder's Brome Grass*).

Although *Bromus Schræderi* is not a native of Britain, and cannot claim to be strictly perennial, it is a valuable forage plant, remarkable for its habit of free growth in early spring and late autumn. The coarse herbage is very sweet, nutritious, and is readily eaten by stock. Constant mowing or grazing is the secret of successful culture, and the growth should not be allowed to attain a greater height than eighteen or twenty inches; four or five crops will then be produced in a year. In warm moist seasons especially its usefulness will be manifested. Several years ago I saw a field of Schræder's Brome Grass which kept an extraordinary head of sheep penned on it. The crop was ready again at one end of the field as soon as the sheep had finished at the other.

The roots feed on the surface, and the plant will thrive on the thinnest soil. So vigorous is the growth that weeds are crowded out, and in my opinion *Bromus Schrœderi* has not been cultivated in England to the extent it deserves.

This is the Prairie Grass of Australia, and is of immense value as a fodder crop in the arid districts of that country. It will therefore be inferred that wherever drought is prevalent, especially for long periods, it is desirable that an experimental plot should be sown, as the plant may prove to be of great service.

CYNOSURUS CRISTATUS

(*Crested Dogstail*).

One of the chief grasses of British agriculture. Associated with Hard Fescue and Sheep's Fescue it may be said to compose the best of our sheep pastures. It constitutes a valuable bottom grass, and is supposed to exercise a beneficial influence in the prevention of foot-rot. Certain it is that sheep show great partiality for the leaves, eating them down so closely that there would be a danger of the plant being exterminated were it not for the fact that the seed-culms soon become hard, and are then rejected by the animals, with the result that seed is matured and shed copiously almost every season. Superficial observers have occasionally misnamed these stalks 'bents,' and have depreciated the grass accordingly. Up to the time of forming seed these stalks do no harm, but the actual production of seed interferes greatly with the aftermath. To prevent this injury it is sometimes worth while to put the mower over the pasture in the early part of July to take off the rising culms.

At the time the crop is cut the herbage of Dogstail is too small to make much difference in the weight of hay, and the fact has been urged as an objection to the sowing of seed. The answer is, that this is distinctly a grazier's grass. In all the most celebrated natural sheep pastures of the country its value for grazing is shown by the extraordinary growth in August, when other grasses are giving up.

On lawns and pleasure-grounds the fine evergreen grass is especially prized; and in deer parks Dogstail should be liberally used, for deer are, if possible, more fond of it than sheep.

Cynosurus thrives in compact soil and will endure conditions under which few other grasses can exist. The roots are hard, and as they penetrate deep the plant is naturally adapted for dry soils, and can withstand drought. Still I have seen it growing luxuriantly on tenacious clays, especially near the sea. Cynosurus forms a conspicuous proportion of the rich pastures in Romney Marsh and other low-lying parts of Kent, and it is remarkable that a grass which does so well in such situations should also find a congenial home on thin upland pastures. On dry loams with a chalk subsoil seed should be sown freely. In fact, there is hardly any soil in the kingdom for which the plant is unsuited. Full maturity is not reached until the second or third year.

The following description of the foliage will aid in determining the presence of Crested Dogstail in pastures. Lower sheaths yellowish white; leaves tapering from base to tip; under surface of leaves glossy; young leaves folded on the mid-rib; ligule truncate.

At Rothamsted and Kidmore *Cynosurus cristatus* appeared to thrive equally well on the unmanured plots and on the plots dressed with mineral manures alone. The plant manifested antipathy for ammonia salts, and an appreciation of nitrate of soda.

For illustration, description, and chemical analysis, see Eighth Edition.

DACTYLIS GLOMERATA

(*Rough Cocksfoot*).

Cocksfoot is indigenous in hedgerows and ditches in almost all parts of England, and is one of the most widely distributed of all the grasses, but prior to the introduction from the United States in the eighteenth century, it does not appear to have been sown for agricultural purposes. The plant shows considerable variations in character in different soils and situations. On dry calcareous land this grass is stunted and wiry, while in fertile valleys and rich land it grows to an immense size. Its proper place is on good, strong, damp soils in low-lying districts, where a great quantity of leafy herbage is produced. In such lands the culms do not change so quickly to woody fibre as on drier soils, and the stalks are eaten down by stock with greater relish. Even on the land which best suits this grass I would sow less of it than is generally advised, and substitute larger proportions of Meadow Fescue and Foxtail. For Cocksfoot is an unsightly grass, growing in great tufts ; the foliage is harsh to the touch, and the coarse hard stems, two or three feet high, are not only objectionable to the eye, but they prevent the hay from fetching top market price, and the flower-heads are extremely liable to ergot. Besides, the aftermath of Cocksfoot is inferior in quality to that of Foxtail, and on thin soils, in a dry season, the former is often very much withered by rust.

It has been asserted that an acre of Cocksfoot will sustain twice as many sheep as an acre of Timothy. This may be true of the early growth, but it certainly does not hold good of the later crop, for the productiveness of the two grasses is nearly reversed after midsummer.

Some years ago a writer gave Cocksfoot greater prominence than older authorities assigned to it. As a result old pastures have unfortunately been ruined by a sowing of the seed, and new pastures have had to be broken up because this grass was sown in too great proportion. Concerning the immense cropping power on strong moist soils there cannot be two opinions; but the plant does not endure long-continued drought so well as Foxtail, and is entirely out of place in upland meadows. When a fine hay crop is the chief desideratum, Cocksfoot should be omitted altogether, and its place be filled by Meadow Fescue.

The valuable qualities of Cocksfoot are even better realised in a three or four years' ley than in a permanent pasture.[1] With Rye Grass and Clover it forms a superior feeding ley, and yields a very bulky crop for consumption on the farm, where coarseness is not objectionable. Full maturity is not attained until the second year.

Cocksfoot comes into flower in June, and meadows in which it abounds must be cut early, so that the flowering culms may be in a young stage of growth. After early cutting, the plant quickly starts again, and while tender the herbage is much relished by all kinds of cattle. Mere size or coarseness is not a standard of excellence, for a smaller crop of Cocksfoot from a rich, light loam has been proved by analysis to be more than equivalent to a heavier crop taken from fen land, and the quality is always higher before flowering than after the seed has been developed. Experiments show that in April ten per cent. of the herbage of this grass consisted of nutritive matter: while at midsummer the proportion was only about three and one-third per cent. These are important facts to be borne in mind in laying down a new pasture. To make Cocksfoot into hay at the end of April, however, would necessitate the sowing

[1] The Swiss authority, Dr. F. G. Stebler, is very emphatic on this point. He says: 'It is chiefly in temporary meadows that it is most advantageous. Yet if sown too largely at first it has the same fault as if sown alone, forming cushions, and consequently an unequal turf. It is better to begin by sowing but little, and, after allowing other sorts to develop properly, to sow a little more later. Only in exceptional cases is it necessary to sow more than 15 per cent. at first. A good rolling is very beneficial in spring. This levels the tufts.'

of this grass alone, because other varieties are not ready for cutting at that time; and Cocksfoot manifests such a strong tendency to grow in clumps, that, with no other grass to occupy the interstices, the soil would be only partially filled with plants.

The great size of Cocksfoot points to its usefulness in covers, although unfortunately it does not answer well with brushwood. Yet it thrives under trees, as is implied by its American name of 'Orchard Grass.' On pastures which are much shaded, a considerable proportion should be sown in company with Foxtail and other grasses which are reliable beneath dripping foliage.

Upon an understocked pasture Cocksfoot will send up flowering stems, and I strongly advocate running the mower over in time to prevent seeding.

The thick, yellow base of the shoot, compressed above, and the strongly keeled leaf, with its erose ligule, will, in the absence of the flowering head, distinguish this species at all seasons of the year.

Cocksfoot when stimulated with ammoniacal manures has a remarkable tendency to smother and starve out the clover of a pasture in which it abounds.

For illustration, description, and chemical analysis, see Eighth Edition.

ELYMUS ARENARIUS

(Sand, or Upright Sea Lyme Grass).

No other grass is equal to this for binding loose or blowing sand, for which purpose the mat-like roots are naturally adapted. Seed has been extensively sown on portions of the English and Dutch coasts. Height, four feet. Flowers in July.

FESTUCA PRATENSIS

(*Meadow Fescue*).

Meadow Fescue is common in Great Britain and on the continent of Europe, and its presence is a fairly certain indication of good land. In situations adapted to the requirements of the grass it scarcely suffers from winter cold or spring frosts. It flourishes in strong, deep soil, especially in low-lying meadows and valleys where a moist atmosphere prevails; and in this country is rarely successful at an elevation of 500 feet above the sea level. *Festuca pratensis* is abundant in the Vale of Aylesbury—noted for the luxuriance of its pastures—and is thoroughly at home in well-drained clay districts. Professor Buckman says: 'In rich meadow flats, as in the Vale of Berkeley, the celebrated locality for the production of Double Gloucester cheese, the *Festuca pratensis* is a common and valuable denizen.' Again he says: 'In an agricultural point of view *Festuca pratensis* is indicated for best lowland pastures not liable to floods.' Scarcely any grass equals this for land under irrigation, although it is entirely unsuited to a water-logged soil.

Some writers have recommended Meadow Fescue for one or two years' leys, but as the plant requires three years to attain maturity, I cannot advise the sowing of seed for a shorter period than three or four years. I regard it as a landlord's grass, to be used principally for permanent pastures.

Festuca pratensis seldom flowers until the end of June, and does not usually mature seed until the first week in August. The ripening of seed lowers the nutritive quality, although it does not greatly diminish the bulk. Sinclair's experiments proved that between the time of flowering and the development of seed the depreciation in feeding value was very serious indeed. But there is no need to anticipate loss on this ground, because, being a late-flowering species, the hay is almost always cut before the danger becomes imminent. One other point deserves mention here. Where hay is cut in the first or second week of June, this grass will not be at its best, and therefore seed should be sown more sparingly than where the hay harvest occurs about midsummer.

The introduction of Meadow Fescue into America furnishes another example of an alteration in character which such a change sometimes brings about. The time of flowering is much earlier there, the plant thrives at a higher altitude, and often grows four feet high, instead of about two feet high as in England. In East Tennessee this grass flourishes and makes superior hay at 1,500 feet above the sea level. The verdant growth during winter in North Carolina has secured for it the local designation of 'Evergreen Grass,' and in Virginia it is known as 'Randall Grass.'

In habit, Meadow Fescue resembles true Perennial Rye Grass, but the herbage of the former is more leafy, and the seed-culms fewer. On light thin soils, and at altitudes where *Festuca pratensis* is unsuitable, an additional quantity of true Perennial Rye Grass, combined with *Festuca duriuscula* and *Festuca ovina tenuifolia*, may well take its place.

Although Meadow Fescue occasionally grows rather rank, the herbage is always tender and succulent, and the quality of the hay first-rate. With cattle the plant is ever a favourite, and, if possible, is even more relished than Foxtail. These two grasses flourish under the same conditions, except that the Fescue scarcely stands heat so well as Foxtail. For practical purposes I am inclined to regard *Festuca pratensis* as the better grass, especially as the germination of the seed is more reliable, and experiments have shown its decided superiority in bulk. Up to the time of ripening seed an acre produced 209 pounds more nutritive matter than did an acre of Foxtail in the whole year. The early growth of Meadow Fescue is not large by comparison, but before the end of June it leaves

Foxtail far behind. After the crop of the latter has passed its prime, the former continues to grow, and thus maintains a supply of herbage which would otherwise be wanting in the pasture. At the time hay is cut Cocksfoot is superior in bulk, but Meadow Fescue largely atones for the deficiency by its increased production afterwards.

Stock show a marked preference for Meadow Fescue over Cocksfoot, if the latter is allowed to grow the least old. The one is invariably eaten down close; the other is frequently permitted to develop into large tufts and send up wiry culms which the animals refuse to touch. Horses also eat *Festuca pratensis* greedily, and seed should always be sown for their benefit on suitable land, especially as the paddocks are often within sight of the mansion, where constant verdure is desirable.

All points considered, this may properly be regarded as one of the most valuable grasses that can be sown.

The following characteristics will help to determine this species in the absence of the flowering panicle: Sheath smooth, reddish purple at base; young leaves involute (rolled) in bud; ligule very short, obtuse, surrounded by a well-marked auricle; blade glossy on lower surface.

At Rothamsted, Meadow Fescue did not show to advantage. It occasionally manifests rather unexpected antipathies as to soil, and in some localities is starved out of pastures by other grasses. Nitrate of soda and mineral manures alone seem capable of augmenting its growth. Stebler, however, speaks favourably of the effects of fresh farmyard manure.

For illustration, description, and chemical analysis, see Eighth Edition.

FESTUCA ELATIOR

(Tall Fescue).

The indigenous variety of *F. elatior*, found in marshes, ditches, and tidal waters of this country, is coarse, harsh, and of little value for good pastures. On imperfectly drained clays, unsuited to the growth of finer grasses, its merits as fodder have been recognised by Sinclair and other authorities. Seed is produced very sparingly and is generally attacked by *clavus* (ergot), which renders it infertile.

The Continental variety, known as *F. arundinacea*, matures fertile seed which is saved and exported to this country. The plant is vigorous in growth and finds a congenial home on moist clays and undrained fens, but is not quite so robust as the native variety, of which specimens may be found in water-meadows of the South of England. The herbage produced by the imported seed is very coarse, but fairly nutritive in quality, either cut for soiling, or after it has been harvested as hay and reduced to chaff.

Experience proves that this Fescue is not permanent on soils which differ materially from its natural habitat, and therefore seed should be sown with judgment, especially as the grass is liable to become ergoted.

True seed when compared with that of *F. pratensis* is distinguishable by its larger size and the more pointed shape of the grain. There is a perceptible roughness on the back of each seed, the three nerves extending the whole length of the grain being armed with minute spines which are visible under an ordinary microscope. In the smaller seed of *F. pratensis* these spines are entirely absent.

The illustration published on page 162 of the Eighth Edition was drawn from a portion of a plant raised from seed of the foreign *F. elatior* (*arundinacea*). It will be observed that the plant is far stronger in habit than is *F. pratensis* illustrated on page 160 of the Eighth Edition. Both were grown alongside in light soil under similar treatment.

When not in flower Tall Fescue may be recognised by the following characteristics of the herbage. Upper surface of the deep green leaf very prominently ribbed, and hard to the touch, especially so at the edges. The round lower sheaths are striated and rough; the bottom of basal sheath purple or red.

For description and chemical analysis, see Eighth Edition.

FESTUCA OVINA

(*Sheep's Fescue*).

Under the name of *Festuca ovina* all the smaller Fescues are occasionally referred to, and much confusion has arisen from the practice. A writer may speak in strong terms against or in favour of *F. ovina*, and it will afterwards appear that he was alluding to *F. duriuscula*. Again, both *F. duriuscula* and *F. rubra*, although perfectly distinct varieties, are not only mistaken for each other, but are also confused with *F. ovina*. The last named possesses certain characteristics by which it may be distinguished from all other small Fescues, and although there is a difficulty in collecting seed it is possible to obtain the true variety; but if true and reliable the price is generally high. As regards luxuriance, colour, and fineness of the herbage, this grass may be termed intermediate between *F. duriuscula* and *F. ovina tenuifolia*, and therefore it will continue to be regarded as more suitable for grazing than for making into hay. On sheep runs and in situations which favour the growth of this Fescue, the succulent foliage will always be appreciated, while for lawns its merits in the formation of a fine close turf cannot fail to claim attention.

FESTUCA OVINA TENUIFOLIA

(Fine-leaved Sheep's Fescue).

Only a small proportion of the seed sold in England under the name of *Festuca ovina tenuifolia* is the true fine-leaved variety. From descriptions which are published in this country it is obvious that the distinguishing characteristics of the seed are unknown to some of those who offer it, and as it generally costs much more than seed of *F. duriuscula* or *F. rubra*, the great diversity in the quotations for this grass is easily accounted for.

The Fine-leaved Sheep's Fescue (*F. ovina tenuifolia*) is the smallest grass cultivated for agricultural purposes, and is perfectly distinct from other Fescues, being densely tufted, with abundance of very narrow leaves and slender flower-stalks. The herbage is succulent, although wiry in appearance, and the culm has the peculiarity of being angular, while all other cultivated Fescues have round stalks. The plant comes into flower the third week of June, grows from six to twelve inches high, and retains its green colour during hot dry weather. It is therefore much used in the making of lawns, for which purpose it possesses a special value in combination with other fine-leaved grasses.

Where Fine-leaved Sheep's Fescue grows freely it is a favourable indication of a healthy neighbourhood.

For hay this grass is practically useless, as the growth is too dwarf to be gathered by machine or scythe. This is pre-eminently a pasture grass for sheep on poor light soils and in hilly country, especially where the subsoil is chalk, and it is generally believed to exercise a beneficial influence on the wool. It forms a principal component of the pastures on chalk downs, where its herbage is very sweet and nourishing, and Southdown mutton doubtless owes its fine flavour and quality to being fed on this wholesome grass. In Scotland it constitutes a large part of the sheep herbage of the Highlands, and all through Russia and Siberia it affords almost the only pasturage for cattle and sheep on dry, sandy, rocky soils where no other grass will grow. In this country horned stock show it no favour.

The roots penetrate to a great depth, and it is noteworthy that as the plant is unable to avail itself of nitrogenous manures, it soon disappears from a pasture which is manured to suit stronger-growing grasses.

Fine-leaved Sheep's Fescue is peculiarly adapted for autumn or, correctly speaking, for late summer sowing. The seedlings are so small that they are liable to be choked by spring weeds, and if sheep are turned in too soon they tear up many plants. But when well established on a poor thin soil *Festuca ovina tenuifolia* maintains its position, and is a match for any intruder. Full development is not attained until the second or third year.

The long, slender, setaceous, folded leaves, no thicker than thread, and tightly encased by the sheaths in pairs and threes, together with the russet green of the foliage, will ensure the recognition of this grass in upland pastures.

On the soils which especially suit Fine-leaved Fescue it is almost indispensable, and the feeding value is very high. For dry hilly pastures a considerable proportion of seed should be included in the mixture.

For illustration, description, and chemical analysis, see Eighth Edition.

FESTUCA DURIUSCULA

(*Hard Fescue*).

The common name applies solely to the flower-heads, which, when ripe, become decidedly hard. The herbage, however, is tender, succulent, and much liked by all kinds of cattle.

Hard Fescue forms one of the principal constituents of the sheep pastures of this country. On all suitable soils it is of importance in forming a close bottom to the turf among stronger-growing varieties, and in this respect is of especial service for upland pastures. Sinclair found Hard Fescue to thrive best in company with *Festuca pratensis* and *Poa trivialis.* His experience is confirmed by my own, and I also find that when sown with Fine-leaved Sheep's Fescue a close and nutritious crop is produced for sheep on down land.

Early in spring Hard Fescue starts into growth, flowering at or a little before midsummer, and yields an abundant lattermath. On moist and rich soils it affords an immense amount of herbage, which sometimes grows higher than the seed-stems, and the plant will endure drought when the land is in good condition. As the nutritive value is much diminished by the formation of seed, the crop should be cut in time to prevent deterioration. In hay the presence of this grass is generally indicative of superior quality, and mountain graziers insist that Hard Fescue contains more 'proof' than many varieties that receive a higher character from the chemist.

Festuca duriuscula is more robust in growth than *F. ovina tenuifolia*, and occupies a position among the smaller Fescues similar to that filled by *F. pratensis* among the taller species. The plant does not grow in tufts, but is perfectly distinct; the seed is larger, and is wanting in the rich golden-brown colour which distinguishes true seed of *Festuca ovina tenuifolia* from that of all other Fescues. On good soils Hard Fescue is also undoubtedly superior to Red Fescue. In dry seasons the former is quite able to hold its own, but in such years the plant develops a partially creeping habit. Red Fescue, on the contrary, in wet summers, loses to some extent its well-known creeping character. This does not make it a matter of indifference as to which of the two is sown. Each variety is constant in character on the soil which suits it, and there is waste of time and loss of fodder in putting either plant on land where it will take several years to adapt itself to uncongenial surroundings.

The folded, almost cylindrical, stiff, deep glaucous green leaves, which radiate from the point of union of the short sheaths, and the absence of a ligule, are sufficient to determine the presence of this species in the pasture.

As Hard Fescue is the most widely distributed of the small Fescues, the seed is naturally the cheapest, and is frequently accepted for the higher-priced sorts by those who are unacquainted with its appearance. This variety may properly be regarded as the least expensive and most desirable base, or bottom-herbage grass, of a permanent mixture for nearly all soils. But for alternate prescriptions it should be used cautiously. On some soils the plant is difficult to eradicate, and after a ley is broken up it is often troublesome among corn.

Stimulating manures do little or nothing for Hard Fescue; they only encourage other grasses at its expense. At Kidmore the bone plots exhibited the most favourable results.

For illustration, description, and chemical analysis, see Eighth Edition.

FESTUCA RUBRA

(*Red Fescue*).

The popular name is derived from the reddish brown colour of the lower leaf-sheaths. The difficulty of obtaining true seed of this grass has fostered the frequently expressed opinion that it is only a form of *F. duriuscula*. But besides a more robust habit, it is further distinguished by its creeping roots, which enable the plant to remain green and succulent when other grasses are burnt up. For this creeping tendency some writers have wholly condemned Red Fescue. I am unable to go that length, and regard the plant as an essential constituent on poor dry harsh soils and upland pastures, especially for its great power of withstanding drought. This quality naturally fits the plant for use on railway slopes, and for all burning soils and hot climates. Yet it endures both cold and shade; indeed, under these conditions the herbage is more abundant. For golf links, putting greens, and recreation grounds Red Fescue is unequalled in the endurance of hard wear. It also offers the further advantage of needing mowing less frequently than some other grasses, and for this reason it is useful on banks and slopes.

Red Fescue shoots rather later in the spring than Fine-leaved Sheep's Fescue, and produces an abundance of small herbage which fills up the bottom of a pasture, and also renders it serviceable in ornamental grounds. The plant flowers in June, and ripens seed about the middle of July. This is one of the few grasses which improve as they get older, the leaves and stems being actually more nutritious, as well as of greater bulk, at the time of ripening seed than earlier in the season. All cattle like the herbage, and with hares it is so great a favourite that a quantity of seed should be sown where this game is preserved in large numbers. Red Fescue must be regarded as especially a pasture grass; for hay it is of small utility, and the lattermath is inconsiderable. Sinclair believed it to attain perfection in the second year, and limited its duration to seven or eight years.

In the absence of the flowering culm, the reddish purple base of the short sheaths, and the long, flexuous, shining, deep green, half-closed leaves, together with the pale-red underground stolons, are a sufficient guide in determining this variety.

The seed resembles that of *Festuca duriuscula*, but is larger and germinates well—decidedly better in the open air than under artificial conditions.

LOLIUM PERENNE

(Perennial Rye Grass).

An American writer enumerates between sixty and seventy varieties of Rye Grass, but no great experience is needed to discover that in so long a list there are more names than sorts. The majority are mere synonyms; others are selections having no permanent character; so that for practical purposes the number of distinct varieties may be reduced to about half a dozen. In the United States, moreover, the term Rye Grass is applied to several species of Elymus; hence it is necessary to make sure as to the species concerned when reading American observations on Rye Grass.

Lolium perenne was the first grass gathered separately for agricultural purposes. It is better known and more frequently used than any other species, and notwithstanding the assertions that it is biennial and not perennial, I am satisfied that it is entitled to the name by which it has been known since 1611, the date of the earliest agricultural book which mentions it. That this grass is not perennial on all soils, nor under adverse conditions, may be freely admitted, and under similar circumstances few, if any, grasses are durable. But when seed of true Perennial Rye Grass is sown on suitable soil and the natural requirements of the plant are met, it will prove to be perpetual.

Perennial Rye Grass was in this country first sown in the Chiltern parts of Oxfordshire, and is still of great service in some portions of that district on cold sour clays, and on light stony land so deficient in lime that it will not grow Sainfoin.

Dr. F. G. Stebler describes it as 'one of the most valuable plants in our meadows. For pasture on clay soils it cannot be replaced by any other plant, and therefore it is largely used in mixtures for forming pasture grass of best quality. In the North of Germany there are even experienced growers who sow only Rye Grass with a little White Clover. *The duration of this plant varies much according to soil and climate.*' I have put the last sentence in italics, because in this remark Dr. Stebler has accurately indicated the origin of the diversity of opinion which prevails concerning the duration of this grass. While Rye Grass is indigenous and perennial in many districts, it stands for a limited time only where soil and climate are unfavourable. On a burning sand or thin gravel it is never permanent, but it answers on a gravelly clay, is at home on all loams, and revels in tenacious land. Even pure clay is not too stiff for it. The poorer and drier the soil, the shorter will be its tenure. On the contrary, the richer and more moist the land, always supposing the drainage to be good, the greater the certainty that Rye Grass will be permanent. The plant responds quickly to irrigation, either of pure water or of liquid manure, but stagnant water soon kills it. The natural habit points to the secret of successful treatment. The roots are maintained by the surface soil, and as poor land speedily becomes exhausted by the free growth, of necessity the plant dies.

Pastures which are stimulated by the droppings of cake-fed cattle, or which are dressed at proper intervals with farmyard manure, continue to grow Rye Grass year after year without re-sowing. When hay is cut early, seeding is, of course, impossible, and if the pasture is grazed the cattle will take care that seed does not ripen. They never permit the heads to flower; the culms and herbage are always closely cropped.

Seed of Perennial Rye Grass germinates with unusual freedom, and it may be relied on to produce a crop under conditions that are fatal to other varieties. The plant matures very quickly and is not injured by the tread of cattle. While other grasses are dependent on season and weather, Rye Grass is able to hold its own under all circumstances, enduring winter frost and summer heat. Another great advantage is that Perennial Rye Grass is so little deteriorated by being allowed to grow old before it is cut. In fact, there is no doubt that as the crop becomes nearly ripe the quality improves, and probably the discrepancies between some chemical analyses and the experience of farmers in feeding stock are traceable to premature cutting.

The high feeding value of Perennial Rye Grass is shown by Dr. J. Augustus Voelcker's report and analysis, from which it will be seen that of the larger grasses Foxtail only is superior to Perennial Rye Grass in nutritive properties. Experiments made by feeding cattle on hay composed exclusively of this grass confirm my estimate of its valuable qualities. It also deserves to be widely known that Rye Grass straw cut into chaff is a very substantial food for cattle. An experienced farmer, probably the largest grower of Rye Grass in this country, informed me that he always prefers to give his store cattle and horses the chaffed straw of Rye Grass, when it has been harvested in good condition, to feeding them on meadow hay cut from an old pasture, and that the animals show a decided preference for the former.

Morton's 'Cyclopædia of Agriculture' contains the following remark on this subject: 'According to M. Péan de Saint-Gilles, a French agriculturist, the ripe straw left after threshing out the seeds is a better fodder than hay made from the green straw. After giving many thousand trusses of the straw to his horses without other food for several months, he found that they ate it as readily as the finest hay, and that it kept them in high condition.'

The objection occasionally urged against Rye Grass, that it does not produce an aftermath, only holds good as to starved crops, and on soils which are not adapted to grow the plant. Even the very cheapness of the seed has been given as a reason for not using it. In the best sense of the term Rye Grass is cheap as compared with many natural grasses, but surely this is a special argument in its favour. It is quite true that *Festuca pratensis* possesses many of its good qualities and excels it in aftermath, but the difference in price is considerable, and the Fescue is distinctly inferior in nutritive qualities, except in its early stage of growth.

One of the main reasons for including Perennial Rye Grass in mixtures for permanent pastures is its reliability, already alluded to, for ensuring a plant. This is a matter of great importance, for if grasses do not fill the ground in the first season, weeds will inevitably appear. And the Rye Grass yields a bulk of hay during the first two years which cannot be obtained without it. Again, Rye Grass fosters the growth of other varieties and aids the general progress and development of such grasses as are slow in attaining maturity. For all these reasons I advocate the use of Perennial Rye Grass seed in prescriptions for permanent pastures. On land which cannot maintain Rye Grass permanently, excellent service will be rendered while it lasts, and the plant will yield up its place when other grasses are sufficiently established to occupy the soil. Meanwhile crops of valuable herbage will be secured.

Most of the Rye Grass seed sown in England is saved in Scotland and in the North of Ireland, and I have no doubt that its acclimatisation in those cold districts tends to maintain its hardiness and its permanency. The popular notion that the first year's crop of Perennial Rye Grass seed produces only an annual plant is a mere fiction, but to ensure a crop consisting exclusively of the true perennial variety it is necessary that maiden seed should be rejected for permanent pasture.

For alternate husbandry Perennial Rye Grass is indispensable, whatever the soil may be. Even on land where the plant certainly would not be permanent, seed should be liberally sown for a short term of years. The excellence of the herbage, the great weight of produce, its early and late growth, and the important fact that it endures the trampling of stock, are all strongly in favour of sowing seed freely.

The endurance of Rye Grass under severe treading deserves more than a mere passing mention. By the sides of country roads a thick formation of turf, the envy of those who wish to make a lawn, is often observable; and this turf will be found to consist almost exclusively of Rye Grass. On one occasion, in company with Mr. Carruthers, we both at the same instant noticed the strong development of Rye Grass at the gate of one of my pastures, then laid up for hay, where cattle usually congregated before milking-time in the seasons when the field was grazed. Examination proved that the Rye Grass was also prominent on the green path across the field. Had the extraordinary growth been near the gate alone, the inference might have been drawn that the increase of such a gross feeder as Rye Grass was attributable to the droppings from cattle kept waiting there; but as the herbage throughout the entire length of the path, which traverses several different soils, is nearly all Rye Grass, it affords evidence of superiority over every other grass as to the capability of the plant for enduring the tread of man and beast. This characteristic accounts for the large proportion of Rye Grass found in the fields of this country which have been depastured from time immemorial.

Exceptionally heavy dressings of nitrogenous manures applied to Rye Grass when in mixture with other varieties for a hay crop are inimical to its existence, because the manures enable coarse grasses to obtain the ascendency. But when Rye Grass is sown alone, or mixed with Broad Clover only, the crop is greatly increased by the application of manures, proving that the plant has no inherent objection to them. Both at Chiswick where Rye Grass was grown alone, and at Rothamsted where it formed only a portion of the herbage, moderate doses of nitrate of soda had a marked effect upon its growth, and the fact was especially noticeable that the nitrate produced an immense amount of root-fibre as well as leafy herbage. It is also interesting to find that at Rothamsted, after decreasing for three years, Rye Grass slightly increased in proportion to the other surrounding grasses, thus affording evidence of its perennial character. Ammonia salts, which proved so powerful a stimulant to some of the stronger grasses, did not benefit Perennial Rye Grass sufficiently to enable it to hold its position against them. On the nitrate plots only was its position maintained. At Kidmore the plant flourished in the bone plots.

The highly glazed under side of the leaves of *L. perenne*, together with the prominent mid-rib, are sufficient to distinguish it at all seasons of the year. The leaf is narrow-tipped, has an obtuse ligule, surrounded by an auricle or collar-like portion of the blade ; whilst the younger leaves are folded throughout their length on the mid-rib when emerging from the flat purple sheath.

For illustration, botanical description, and chemical analysis of Sutton's Perennial Rye Grass—a branching variety, considerably dwarfer than Pacey's, and still more perennial in character—see Eighth Edition.

LOLIUM VULGARE. L. ANNUUM

(*Common*, or *Annual Rye Grass*).

Seed of this plant is larger and flatter than that of the Perennial and Italian varieties. It is almost as broad as the seed of ***Bromus mollis***, but is lighter and awnless. Annual Rye Grass deserves to be better known, for it possesses valuable properties, yielding a good bulk of nourishing herbage of a light green colour, and sending up a dense mass of flowering culms. The roots, being small, are more easily got rid of when the ley is ploughed than are the roots of the perennial variety; and there is a general concurrence of opinion that although the plant is of such rapid growth, and draws its nourishment entirely from the surface soil, it does not greatly impoverish the land for the next crop.

The objections which have been urged against the use of Perennial Rye Grass in permanent mixtures are largely owing to the substitution of Annual Rye Grass. The latter is only biennial in duration, and there are several reasons why seed of so short-lived a plant should not be included in prescriptions for permanent pasture.

LOLIUM ITALICUM

(Italian Rye Grass).

I wish to prevent the possibility of a misunderstanding by saying at once that this grass is under consideration here solely in its proper connection with alternate husbandry. In a permanent pasture it is generally harmful, and the seed should rarely be included in a prescription for that purpose.

Italian Rye Grass was probably first cultivated in Lombardy, and spread thence through Europe. It was introduced into England in 1831 by the late Mr. Charles Lawson, of Edinburgh, who was an old friend of my father. At first Mr. Lawson imported seed from Hamburg, but a few years later he was able to obtain it direct from Leghorn. The experiments made with Italian Rye Grass by the late Mr. William Dickinson resulted in crops which were then considered marvellous, and for many years his name was prominently associated with this fodder plant. His system no doubt suggested the idea of sewage treatment, and it is not too much to say that without Italian Rye Grass the present method of sewage farming could not be carried on. There is no other green crop, except perhaps *Bromus inermis*, which can be substituted for it, or which will yield equally valuable herbage.

There are considerable variations in the character, productiveness, and duration of the several strains of this plant; still, roughly speaking, all are biennial. The seed which is imported from the Continent yields very heavy crops, but is so infested with pernicious weeds as to need most careful cleaning before it is fit to sow. To avoid the risk of making the land foul, seed has long been grown in Ireland, and continuous selection has developed all the good qualities of the imported stock, except its extreme vigour. *Lolium italicum Suttoni,* which is now largely grown, was obtained by careful selection of the most free-growing and leafy plants. This stock grows with extraordinary rapidity, with but a small proportion of stalks, so that the crop, whether green or made into hay, is of specially fine quality.

Italian Rye Grass will grow in almost any soil, but is least satisfactory on poor dry land, unless it can be freely assisted with liquid manure. Fair results have, however, been obtained from heaths dressed with marl and farmyard dung. This grass needs warmth and moisture, and in rich damp soils the growth is extremely rapid. Irrigation by liquid manure results in enormous crops, following each other in surprisingly rapid succession; and the plant is so hardy as to endure our coldest winters without injury, while it starts earlier in spring and grows later in autumn than any other grass.

Italian Rye Grass is so much preferred by stock, that when a two years' ley which contains it is fed off, the cattle will not allow a single flower-stem to ripen. The forage promotes a great flow of milk, and improves the flavour of butter and cheese. The celebrated Parmesan cheese is said to be made from the milk of cows fed entirely on Italian Rye Grass.

No wonder, therefore, that with such remarkable qualities the plant should have been recommended and used for permanent pastures, although nearly every authority has pronounced it unsuitable for that purpose. Italian Rye Grass is so gross a feeder as actually to choke and smother the Poas and finer Fescues, instead of simply nursing and sheltering them from scorching heat and severe frost as does the comparatively slower-growing Perennial Rye Grass. Valuable as Italian is for alternate husbandry in the company of other strong-growing grasses, such as Cocksfoot and Timothy, I have always objected to its use in permanent pastures, even during the years when the contrary custom prevailed. This view is now admitted to be sound, and the circumstances must be altogether exceptional which warrant the admixture of any proportion of the seed in a prescription for permanent pasture.

With singular inconsistency, some writers who disapprove of Italian Rye Grass in a mixture of permanent grasses, advocate its employment for renovating an old pasture. A more illogical proposition has never been made. When the brief existence of Italian Rye Grass terminates, the pasture will be left in a worse state than before.

There is a prejudice against this grass because it severely taxes the soil, and it is quite true that Italian Rye Grass will appropriate all the nourishment it can get. But this is an argument in its favour. It means that the plant is a powerful agent in extracting constituents from the soil, which it yields up in valuable food for cattle. Given the desire of obtaining from an acre of liberally fed land the largest possible produce within twelve months, and Italian Rye Grass has no equal for fulfilling the condition.

Another prejudice is that Italian Rye Grass is a bad preparation for wheat. On that point Mr. Dickinson's own words can be quoted: 'Thirty sheep may be kept upon Italian Rye Grass, fed through hurdles, upon as little land as ten can be kept upon the common system upon common grass, and the finest crops of wheat, barley, oats, and beans may be grown after the Italian Rye Grass has been fed off the two years of its existence. Wheat invariably follows the Italian, and splendid crops are grown.'

Again, it is objected that twitch is produced by Italian Rye Grass. With pure seed this is impossible, but plenty of Italian Rye Grass seed infested with twitch is freely bought and sold every year. The remedy is in the buyer's own hands. When twitch is already in the soil, a starved crop of Rye Grass may give the couch an opportunity of asserting itself. Still, the tendency of a well-fed and frequently cut crop of Italian always favours the destruction of twitch.

Italian Rye Grass is very frequently sown with Broad Clover in composite mixtures for alternate husbandry, and it may also be used with *Avena elatior* when a more durable crop is wanted. But the most profitable method is to grow Italian Rye Grass alone, because the crop can be cut before the stalks become hard and lose their nutritive qualities. Another reason why Italian should not be allowed to grow old is that the plant is extremely succulent, and in warm thundery weather there is a risk of the roots beginning to rot. On this account a very bulky crop cannot be safely left for seed in a hot, damp summer.

The usual months for sowing are February, March, September, and October, and the quantity three bushels per acre, but seed may be sown at any time from spring to autumn. The plant is rolled in spring, the crop is cut frequently, and the land broken up in the following spring twelvemonth. Many farmers make it a rule to sow in October, and on warm moist soils crops have sometimes been cut at Christmas, and again in the following April. Such experience, however, is exceptional, but it proves the hardiness and fertility of the plant.

Italian Rye Grass, when out of flower, may be recognised by the same characteristics which distinguish *L. perenne*, with this difference: in *L. perenne* the sheath is flat, while in *L. italicum* the sheath is round.

For illustration, description, and chemical analysis, see Eighth Edition.

PHLEUM PRATENSE

(*Timothy*, or *Meadow Catstail*).

To see this grass in perfection a visit must be paid to the United States, where it is grown alone, and yields amazing crops of hay, three, four, and in one recorded instance six feet high. In the pastures of that country some of our finer meadow grasses are unknown and Timothy is more highly prized than with us. It is supposed to have been introduced into England for agricultural purposes about the middle of the eighteenth century, and is now widely distributed. The plant is perennial and fibrous-rooted, but under certain conditions the root shows a tendency to take a bulbous form. Indeed, the character of the plant varies considerably in different soils and situations.

Timothy luxuriates in clay districts and on moist soils, and on peat the growth is unsurpassed. This is one of the grasses most certain to grow, and is specially serviceable in the company of Foxtail, because it fills the interval between the first growth and the aftermath of that grass. Another substantial advantage is that Timothy attains the height of its productiveness in the first year after being sown, but unless it is properly fed there is a tendency to weakness in the third or fourth year; and considering the tax which it necessarily makes on the soil, this will occasion no surprise.

No other grass will bear extremes of heat and cold better than Timothy, and in rigorous winters, which severely test the endurance of other hardy grasses, this will stand the trial uninjured. All cattle eat the herbage greedily, and horses manifest especial fondness for it. Timothy adds materially to the weight of the hay crop, and is not only highly nutritious while young, but the feeding value increases until the seed is formed. If, however, the plant is allowed to mature seed, little or no aftermath is to be expected.

This grass is generally regarded as a late variety, and correctly so, because it flowers in July. Still, the early herbage is good, and sheep may be run over a pasture containing a large proportion of it until late in spring without endangering the crop of hay. The plant is really mown to greatest advantage before the ears are out of their sheaths, having regard to the quality of the hay and the quantity of the aftermath. It should be clearly understood that I am here alluding to the treatment of a meadow containing an unusual proportion of this grass.

When grown alone as a fodder plant, Timothy may be allowed to remain uncut not only until in full flower, but even for ten days or a fortnight later, and during that time the feeding value will be augmented. Sinclair says that, 'subject to the weather being such as to keep it in growth, every two days' growth after flowering will increase its nutritive value as much as any eight days' growth before that period.' But these hard wiry stalks, when dried, can scarcely be called hay, and although they may be nutritious, it is desirable to reduce them to chaff before they are given to cattle.

Timothy should form a constituent of every permanent prescription for heavy soils, but as in recent years there has been a tendency to give undue prominence to this grass, I consider it necessary to say that an excessive proportion in a pasture may be mischievous. The herbage is coarse, the stalks soon become hard, and in that state the increased feeding value is of no service if cattle refuse to graze them, as they undoubtedly do; or if the abundant presence of this grass reduces the price of the hay.

When seed is freely shed there is a danger that the plant will take almost exclusive possession of the land, especially on those formations which favour its growth. And once sown it is a very difficult grass to eradicate. All points considered, great judgment is requisite in order to realise the full advantage of Timothy, and at the same time to maintain the general efficiency of the pasture.

A marked characteristic of the plant is its extreme hardiness, and for this quality Timothy is extensively grown in Canada in association with Alsike Clover, for one and two years' leys. For the same purpose it has won favour in Scotland, where experience has proved that hay made from Timothy and Alsike is highly sustaining food for horses in hard work. Such hay, when well harvested, is in constant demand at a profitable figure. Another excellent crop is obtained by sowing a mixture of Timothy and Cow Grass. It would perhaps be impossible to find two plants which are more suitable for growing together. For alternate husbandry generally Timothy offers very substantial benefits ; the only objection to its employment for a term of years is the trouble experienced in getting rid of it when the ley has to be broken up.

The greyish-green tint of the short, broad leaves, and involute character of the emerging leaf, together with the smooth sheath and obtuse ligule, clearly distinguish this grass from others when the cylindrical flower-spike is wanting.

For illustration, description, and chemical analysis, see Eighth Edition.

POA PRATENSIS

(*Smooth-stalked Meadow Grass*).

Although there is a general resemblance between this grass and *Poa trivialis,* the two plants differ materially in habit, character, and structure. *Poa pratensis* is unlike *Poa trivialis* in having broader and blunter foliage, an obtuse ligule to its leaf, smooth stalks, and creeping roots that send out long stolons which take a horizontal direction and form little tufts. Each species has a value of its own, and is fitted to serve a purpose for which the other is less adapted. *Poa pratensis* is naturally suitable for enduring drought, while

Poa trivialis thrives in moist land. This is the chief constitutional difference between the two plants, and although each of them will to some extent conform to the conditions which specially favour the other, yet the distinction clearly indicates the use to which each one should, as a rule, be applied.

Poa pratensis does not root very deeply, but is dependent principally upon the surface soil, and is therefore met with on all geological formations if the surface happens to be suitable. All eminent authorities concur in recommending it for good dry soils, and my own experience has convinced me of the correctness of their opinion, but light land must be rich in humus to ensure success. On heavy and tenacious soils this grass will often grow, but it is not then seen to the greatest advantage.

During the first year *Poa pratensis* remains small and does not throw up any stalks. When fully established it flowers once annually; after the stems are cut, no others are formed until the following season. Primarily this is a pasture grass, because the flowers rise too early for the scythe. Otherwise, for very forward cutting it is valuable for hay, and when the crop has been taken, a good aftermath of leafy herbage follows. The earliness of *Poa pratensis* is no doubt attributable to the fact that the plant appears to be insensible to cold, or growth would not continue during spring frosts. The presence of a large proportion of this grass will, by the shelter it affords to other varieties, make an early pasture, and for this reason it is of especial service to the grazier, to whom half a ton of feed at the end of March or the beginning of April is of more importance than a ton in May.

For lawns, Smooth-stalked Meadow Grass may be freely sown, as it is strictly perennial, and forms a fine compact turf.

In the West of Ireland, where this grass grows all the year round, it should be largely used; and, as one of the most suitable grasses for fogging, it is serviceable in Wales. For railway embankments or other situations on which a rapid matting turf is wanted for holding the soil together, no more valuable plant can be introduced.

Under the name of Kentucky Blue Grass, or June Grass, *Poa pratensis* has been cultivated in the United States with striking success. When once established as a pasture the value of the land is immediately enhanced. Americans grow it on their richest soil, and are surprised that we use the plant so little and never sow it alone.

Experience of a different character has been realised in some parts of New Zealand, where this grass has developed a very objectionable habit of growth and is difficult to eradicate from arable land when once it has obtained a footing. This fact does not prove the worthlessness of the grass for British use. The case is analogous to that of our familiar Sweetbriar, which since its introduction into Tasmania forms thickets in that country, entangling and overpowering other vegetation, throwing underground suckers across fields, and matting the soil with a dense mass of fibre. But we are not on that account inclined to sacrifice the charming bush of our gardens and hedgerows. Neither the merits of *Poa pratensis* when cultivated in America, nor its demerits when sown in certain districts of New Zealand, however interesting they may be, need influence our estimate of the value of a grass which experience has proved to be of great service in this country.

The rounded, concave tip of the leaf, together with the obtuse ligule, the yellowish green of the lower sheaths, and short stolons, will enable the observer to determine the presence of this species in a pasture.

Dressings of nitrate of soda discourage the growth of *Poa pratensis*, while mineral superphosphates combined with ammonia salts foster it. At Kidmore decorticated cotton cake was also conducive to the growth of this grass.

For illustration, description, and chemical analysis, see Eighth Edition.

POA TRIVIALIS

(*Rough-stalked Meadow Grass,* or *Orcheston Grass*).

This variety differs from *Poa pratensis* in having narrower foliage, a long pointed instead of a blunt ligule to its leaf, fibrous instead of creeping roots, and slightly rough sheaths. The roughness is only apparent to a sensitive touch, but is readily perceived when the grass is drawn across the tongue.

Poa trivialis was first sown in this country about 1780, and in recent years it has been mistakenly recommended for all purposes. On strong moist soils, where *Poa trivialis* is quite at home, it surpasses *Poa pratensis* both in quantity and in quality, but the favourable comparison does not extend to light land liable to burn, on which the superiority of *Poa pratensis* becomes manifest. The plant is subject to injury by spring and autumn frosts, and also by severe winters. It is rich in potash and phosphoric acid, and soon exhausts land which is not naturally good or constantly fed.

This grass forms a large part of the turf of valleys in mountainous districts where the rainfall is considerable. It thrives under trees, prospers in water-meadows and all moist situations, and bears well the hoofs of stock, but it must be mown before the foliage turns yellow at the base, or it is liable to rot. Sinclair says that it is unprofitable for any purpose on dry exposed situations.

Poa trivialis flowers at the usual time for cutting hay, but is especially useful for pastures nevertheless. The herbage is of more value at the time of ripening seed than earlier in the season, but as ripening does not take place until the end of July, it is impossible in the South of England to keep the hay crop waiting for it. The aftermath is good after early cutting only. Sinclair advises its use for permanent pastures on rich soils and in sheltered positions, and I quite agree with this view. Valuable as this grass is for such situations, I do not consider that for any other land it is worth while to incur the cost of seed, which, if true, is always expensive. Wherever *Poa pratensis* will answer, it should be used instead.

When not in flower this species is recognisable by the slender leaf, which gradually tapers from the base to the tip of the blade, and has a well-marked keel. The young leaves are compressed, and folded on the mid-rib, the lower sheaths are loose and tinged violet-purple.

At Kidmore nitrate of soda and muriate of potash had a marked effect on the development of this grass; ammonia salts diminished its growth. The Rothamsted experiments have clearly established the fact that in this respect it differs altogether from *Poa pratensis.*

For illustration, description, and chemical analysis, see Eighth Edition.

POA NEMORALIS

(*Wood Meadow Grass*).

Wood Meadow Grass starts growth very early in spring, yields a greater bulk of herbage than *Poa pratensis,* bears drought remarkably well, and is the most valuable of the cultivated Poas. For lawns and ornamental grounds it answers admirably, and is indispensable for sowing under the shade of trees. The seed is too costly, and the supply too uncertain, to warrant a large use of it in ordinary grass-land farming.

The very short sheath—not more than one-quarter of the length of the flat blade—together with the laciniated obtuse ligule, and the compressed character of the younger leaves, are points by which this variety may be recognised when not in flower.

For illustration, description, and chemical analysis, see Eighth Edition.

POA SEROTINA

(*Late Meadow Grass*).

In comparison with *P. pratensis* and *P. trivialis* this variety is later in flowering and in ripening seed. The plant is successful on almost all kinds of soil, especially on rich moist land. For hay it is a valuable grass, producing an abundance of foliage early in the season. The aftermath is not very great, but in permanent pastures it has the merit of growing freely between the period of ripening and after-yield of most other grasses.

Quite a year is required to develop strong plants of Late Meadow Grass, and for this reason it cannot be recommended for alternate husbandry. True seed is difficult to obtain.

In the absence of the flowering panicle *Poa serotina* can always be distinguished from *Poa nemoralis* by the short blunt ligule.

POA AQUATICA

(*Water-Meadow* or *Sweet Reed Grass*).

Adapted for low-lying land subject to floods, nutritious, and generally liked by cattle. In the Fen districts it forms a large part of the herbage, and yields abundant crops of hay. From the name it will be understood that this grass is specially suitable for water-meadows, and seed may also be included in permanent mixtures for undrained clay. The flowering occurs in July and August.

Poa aquatica is also known as *Glyceria aquatica*, and is common in the water-meadows of the South of England, where the closely allied Floating Sweet Grass, *Glyceria fluitans*, is also frequently found.

Agricultural Clovers
(Leguminosæ)

TRIFOLIUM REPENS PERENNE
(Perennial White Clover).

This plant was first cultivated in the Netherlands; hence the familiar name, Dutch Clover. Probably it was not sown in England until the beginning of the eighteenth century, although it is now indigenous all over the country. The seed will lie dormant for a long time and at a great depth, and be ready to spring into life when brought to the surface. The habit of the plant is creeping, and when once established it soon covers the ground. Sometimes its luxuriance is excessive, and the numerous flower-heads, for which cattle manifest no partiality, are a disadvantage, so that in sowing White Clover judgment should be exercised. In warm, rainy seasons it spreads rapidly, but makes little or no progress during cold, dry weather. Besides the mass of fibrous surface roots there is a long tap-root which goes deep into the subsoil, sustaining the plant during drought, when only the parent stem grows, the lateral and creeping shoots remaining dormant.

The character of this clover differs materially, according to the soil on which it is grown. Sinclair remarks that 'it maintains itself in soils of opposite natures,' because of its peculiarity in having two distinct forms of root. It prospers on mellow land containing lime, and on all soils rich in humus, from marl to gravel, or gravelly clay. In poor land it does better, and is less sensitive to atmospheric influences, than Red Clover, and is of great importance on land which cannot be depended on to grow Perennial Red Clover. In early spring it produces very little food, and the plant is so dwarf that it is practically useless for cutting. For a crop of hay Alsike is preferable. Still, Perennial White Clover forms an essential constituent of every good pasture. All cattle eat the herbage with relish, but it is less useful for the production of milk than of flesh, and is of special service in fattening sheep. No doubt the crop is more palatable to stock before the plant flowers than afterwards; indeed, a profusion of flowers is no indication of an abundant bite. White Clover is not suitable for culture alone, and it is better for cattle when mingled with grasses, especially with Perennial Rye Grass. The Norfolk farmers largely use it for ewes and lambs, but from difference of climate the favourable opinion they entertain is not shared by practical men in the West of England.

The fertilisation of White Clover is aided by insects. From ten flower-heads visited by bees Darwin obtained ten times as many seeds as from a corresponding number protected by gauze. On a subsequent occasion he failed to obtain a single fertile seed from twenty protected heads.

When White Clover is one of the constituents of mixed grasses and clovers, the application of ammonia salts and nitrate of soda so develops the gramineous herbage that there is a smaller growth of White Clover than when the land is left unmanured. Mineral manures—basic slag especially—considerably augment the crop of White Clover, and a similar result is obtained from dressings of marl or vegetable ash.

For illustration, description, and chemical analysis, see Eighth Edition.

TRIFOLIUM PRATENSE

(*Red*, or *Broad Clover*).

Red Clover is said to be indigenous in every country in Europe except Greece. In a wild state its presence is an indication of the fertility of the soil. Although a strong-land plant, it will grow on almost any soil, and contains so much moisture that only one-fifth of the weight of the green crop is found in the haystack.

Winter and spring frosts are very injurious to Red Clover, and to save the plant a top-dressing of long manure is sometimes necessary, for which, however, there is a return in due time.

A considerable diversity can be discerned in the various strains of Red Clover. Seed is imported from all parts of Europe, and large quantities from North and South America. Each country has one or two well-defined types of this plant, and although growers may be careful to avoid buying any but so-called English seed, the fact that the stock may only have been imported two years previously will account for the differences which are every year visible in crops of Red Clover. The prejudice existing against foreign seed, especially that from America and France, is well founded. Experience has proved that seed from either country produces a smaller crop than can be obtained from a stock which has been acclimatised in England for many years, and there is also the great danger of Dodder to be considered. The most virulent form of this parasite is found in the Red Clover imported from Chili. Seed from that country is infested with *Cuscuta chilensis*, a species of Dodder so large in the grain as to render it most difficult to eliminate from the clover seed. So far as European seed is concerned, there can be no doubt that the Broad Clover grown in Styria and in some districts of North Germany is as robust and hardy as English stocks.

Trifolium pratense is quite unsuited for permanent pastures, but should form a large proportion of an alternate mixture. The great root-growth made during the two years of its existence is the best possible preparation for the following wheat crop.

Red Clover is incapable of self-fertilisation, and the Humble Bee is almost exclusively the medium by which pollen is conveyed from anther to stigma. In Australia and New Zealand, until the Humble Bee was introduced, seed was rarely ripened, and the Red Clover sown in those colonies was all imported, principally from England.

The character and culture of Red Clover are so well understood as to render further remark needless.

For illustration, description, and chemical analysis, see Eighth Edition.

TRIFOLIUM PRATENSE PERENNE

(*Perennial Red Clover,* or *Cow Grass*).

This clover doubtless originated in a cross between *Trifolium pratense,* or Broad Clover, and *Trifolium medium,* or Zigzag Clover.[1] The latter has never been in commerce, nor has it been grown as a crop, except for experimental purposes. Yet some writers have fallen into the error of confusing it with the true Perennial Red Clover, which in Berkshire, Oxfordshire, Hampshire, and Wiltshire is known as Cow Grass. In these counties Cow Grass is of immense importance, and enters largely into the rotation of arable land. The use of it is extending to other parts of England,

[1] *Trifolium medium.*—The Cow Grass, or Zigzag Clover, of botanists is so called from the decided zigzag growth. So distinct is the plant that it can scarcely be classed with Red Clover at all. The blossom is darker in colour than that of *Trifolium pratense,* the head less dense, invariably growing on a long peduncle instead of immediately adjacent to the leaf. The absence of the broad membranous stipule, and the substitution of one which is long and narrow, terminating gradually in a spear-like point, is also a marked characteristic, by which those who have once observed this peculiarity are never again likely to mistake *Trifolium medium* for *Trifolium pratense.*

and it is astonishing that so valuable a plant has not long since been widely recognised as indispensable for temporary pastures, instead of being restricted to permanent pastures only.

Trifolium pratense perenne differs from Broad Clover in having a somewhat taller, smoother, and, except in its very young state, a less hairy stem, and a stronger, less fibrous, and more penetrating root. The plant carries its flowers some way above the foliage, surpasses Broad Clover in succulence and weight of crop, and stands frosts much better.

The root of Perennial Red Clover reaches down into the subsoil, enabling it to obtain moisture and nourishment in the hottest weather, when Red Clover gives up from drought. This penetrating habit also affords a means of sustenance to the plant on land which is too poor to grow Broad Clover, and frequently makes it desirable to increase the proportion of this seed for pastures on thin uplands.

Perennial Red Clover has two characteristics which greatly augment its value : flowering does not begin until at least ten days later than Broad Clover, and the more robust and solid stems remain succulent and eatable by stock long after Broad Clover has become pithy and withered. Perennial Red Clover fills up the gap between the first and second cuttings of Broad Clover, and comes into use at a time when no other green food is available for the horses of the farm, but it rarely gives a second crop of any consequence.

Cow Grass produces comparatively little seed from its single crop; whereas Red Clover yields a good crop of seed from the second cutting, after the first has been taken as fodder. Consequently seed of the perennial variety is necessarily high in price.

It is worth noting that Cow Grass is understood on Mark Lane, and in many parts of England, to mean no more than a fine handsome sample of Broad Red Clover. The true Perennial Red Clover is rarely obtainable, except from those who make its culture a study. And, just as all Rye Grass has been condemned because the annual variety has been used where only the perennial kind should have been employed, so true Cow Grass—*Trifolium pratense perenne*—has been disparaged because Broad Clover has been called by its name and supplied as the genuine article.

When the late Mr. H. M. Jenkins was in Flanders on his tour with the late Mr. James Howard some years ago, he found that this Perennial Clover was relied on as the chief soiling crop, and was used in exactly the same way as in the English counties I have named. So far as I am aware, however, true Belgian Cow Grass is never offered in this country, and if it were offered, there would be the risk of its being foul.

The true Perennial Red Clover is an invaluable plant for permanent pastures, and should be included in every mixture for that purpose. Its presence in a pasture at midsummer, when Alsike is giving up, is of great service, and although it does not produce a second crop for the scythe, it yields a quantity of excellent food. Perennial Red Clover is not perpetuated in pastures by seed, as is commonly supposed, but from short stout branches extended from the parent plant, which root and take the place of the parent should conditions of soil or climate interfere with its perennial character.

Dr. Stebler observed that the plant differs from ordinary Red Clover in having a less fibrous root, stalks generally solid instead of hollow, and that it produces fewer flowers, and therefore a smaller yield of seed. As a consequence true seed is always expensive and difficult to obtain. Dr. Stebler also supports my opinion that this clover is more strictly perennial than any other variety, and he distinctly states that it is a mistake to confound *Trifolium pratense perenne* with *Trifolium medium.*

At Rothamsted, ammonia salts had the effect of eliminating this plant from the various plots to which they were applied, whether alone or in conjunction with mineral manures. Nitrate of soda also diminished the growth. Even potash and mineral manures did not maintain the permanence of this clover. It is a remarkable fact that the unmanured plots, where there was little other herbage to interfere with it, were the only plots on which Cow Grass retained its original position. The Rothamsted experiments, however, decisively proved this plant to be more enduring than *Trifolium repens*; hence there can be no doubt whatever as to its great value in laying down land to pasture.

For illustration, description, and chemical analysis, see Eighth Edition.

TRIFOLIUM HYBRIDUM

(*Alsike Clover*).

Although this plant was named by Linnæus, there is reason to doubt whether it is a hybrid between Red and White Clover, as is generally assumed. The common name of Alsike is derived from the village of Syke, near Upsala in Sweden, where it has been grown with great success. The plant is indigenous in Southern Europe, yet it does not appear to have been cultivated until after the example was set in Sweden, whence it reached this country.

Alsike Clover is a true perennial, and on the Greensand formation comes up year after year where once it has been sown, but on some light soils it is not permanent. Were it not for the surface growth, and the consequent inability to withstand drought, Alsike would take a more important position in agricultural operations than Broad Clover, especially as the former will often grow on clover-sick land.

The plant is peculiarly adapted for damp soils, and is decidedly more productive in wet than in dry seasons; it endures heat and cold well, succeeds in undrained clays better than any other variety, and is the only clover that will stand irrigation. But in the event of a crop being laid on a damp soil it should be promptly cut; if left it may begin to rot at the base.

For meadows and pastures Alsike Clover is far superior to White Clover, as it produces a crop tall enough for the scythe, and materially adds to the total weight of hay. It also produces more keep and is better relished than White Clover. On some soils Alsike will grow quite as much herbage as Red Clover, and as the vegetation is principally composed of leaves, care is requisite in the process of making hay, to prevent them from being broken off and lost, especially as the plant is more succulent than Red Clover and takes longer to make into hay.

The time of flowering approximates to that of Cow Grass, and the nutritive value is highest at the flowering stage. Alsike is not hasty in getting old and pithy, and renews growth quickly after cutting, although the total of the aftermath is only equal to half the first cut. Still, the plant has a high value for pastures, and is one of the best for alternate husbandry. It flourishes in the same deep moist soil as Timothy, and makes an excellent companion to that grass. With Tall Oat Grass, Italian Rye Grass, and Cocksfoot it also combines well and yields excellent forage.

For illustration, description, and chemical analysis, see Eighth Edition.

TRIFOLIUM MINUS

(*Yellow Suckling*).

Notwithstanding the fact that this is an annual clover and the growth is very small, yielding but scanty produce on the gravelly soils and stony land where Yellow Suckling finds a congenial home, it is by no means unworthy of a place in a permanent mixture. On the Greensand this plant is of considerable value, forming a dense mass of herbage and seeding itself down every year. When grown with Rye Grass, and cut before the herbage is too old, it makes excellent hay which is much relished by stock. In such a case the roots cannot be depended on for another crop, but if pastured and kept constantly down the plant will provide a bite all the summer.

The small foliage of *Trifolium minus* is admirably suited for lawns and pleasure grounds.

MEDICAGO LUPULINA

(*Common Yellow Clover, Trefoil, Nonsuch, Black Grass,* or '*Hop*').

A fibrous-rooted biennial which flowers from May to August, and to a casual observer closely resembles Yellow Suckling; but the foliage is a paler green, the stems much less hard and wiry, the flowers not so dark, whilst the entire plant is usually covered with a fine down. The black seed-pods, which have earned for it the name of 'Black Medick,' are also useful as a means of identification. Trefoil starts so early in spring as to give a bite before any other clover, and flowers ten or fifteen days in advance of *Trifolium pratense*.

Although the procumbent habit of growth prevents the plant from yielding much herbage for the scythe, and there is very little aftermath, the quantity of keep in spring is considerable. Trefoil is primarily adapted for alternate husbandry, being only biennial in duration. Seed is, however, so freely shed by this clover as to render it practically permanent, and as chemical analysis shows the herbage to be nutritious, there is good reason for including a small proportion in most permanent mixtures, especially as it grows compactly and helps to make a good bottom to a pasture. Where Trefoil is indigenous in abundance, seed need not be sown, nor should it be admitted to those soils on which it is liable to smother other plants.

Common Yellow Clover endures cold better than heat, will grow freely on almost any soil, and shows preference for such as are calcareous, because lime is one of its necessities. Clay marl is its special home. Fresh green manure does not suit the plant, but a top-dressing of a good compost or of vegetable ash generally brings a full crop. All manures containing potash and phosphoric acid have a magical effect on its growth. Sheep injure it less than they do Red Clover.

Trefoil is rarely sown alone, but frequently with White Clover, and it is important that the Trefoil should not be allowed to predominate, or it may take entire possession of the land. Foxtail, Sweet Vernal, and Smooth-stalked Meadow Grass combine well with Trefoil in affording early and valuable herbage.

Many farmers call this clover 'Hop,' but it must not be confounded with the true Hop Clover, *Trifolium procumbens*, a native British plant, which is not used in agriculture.

For illustration, description, and chemical analysis, see Eighth Edition.

LOTUS CORNICULATUS

(*Birdsfoot Trefoil*).

As a strict perennial, attaining maturity in the second year after sowing, this variety often fills up the bottom of a pasture when other sorts are deficient, and will do this on land which is clover-sick. Birdsfoot Trefoil is useful for all soils, and is remarkable for its endurance of drought, however extreme. On high exposed thin soils it is more productive than any other clover, and it augments the total of the crop without apparent detriment to the taller, upright-growing varieties. Therefore, if expense need not be considered, seed in limited quantity should be sown. Unfortunately, such soils do not often warrant the outlay for this clover, the price of which is almost prohibitory.

Birdsfoot Trefoil is eaten with avidity by all kinds of stock. The plant has narrow leaves, a smooth stem, and bright yellow flowers, occasionally tinged with red.

SUNDRY PLANTS USED FOR GRAZING, FEEDING, OR MAKING INTO HAY

ACHILLEA MILLEFOLIUM

(*Yarrow*, or *Milfoil*).

YARROW is neither a grass nor a clover, but a compositaceous plant, like the daisy, the tansy, and the chamomile. As an ever-present constituent of dry pastures it naturally claims consideration here. The leaves are dense, elegantly fimbriated, and the flowering time is after midsummer. The plant is strictly perennial, and multiplies itself by root-growth. It appears to be independent of the presence of potash in the soil, and flourishes where clovers scarcely maintain existence or fail entirely.

Achillea Millefolium is adapted for pastures, not for meadows, and seed should be excluded from any mixture for a crop that is to be generally mown. Sheep are very fond of the herbage, which is believed to impart an excellent flavour to mutton, venison, and to butter also, notwithstanding its astringent qualities.

On dry gravels and light sands that will scarcely support any other plant Yarrow will grow, and no summer is too hot for it. So great is its power of enduring drought that after a dry season there is always a large demand for seed, and as this is difficult to obtain of reliable germination, the price often runs up to an extravagant figure.

Ammonia salts appear to be inimical to the growth of this plant, nitrate less so, but heavy doses of any ammoniacal manure are certain to reduce its bulk by increasing the strength of surrounding grasses. Mineral manures alone seem to favour its development.

For illustration, description, and chemical analysis, see Eighth Edition.

CICHORIUM INTYBUS

(*Chicory*, or *Succory*).

A perennial which grows wild in dry wastes, on roadsides, and the borders of fields in many parts of England. When in bloom its bright blue flowers are very conspicuous. The plant is cultivated to a considerable extent on the Continent, and the blanched leaves have been used from time immemorial for salads. Chicory flourishes on nearly every class of soil, and has been proved to be profitable on poor sands as well as on richer and more productive land. Associated with Burnet and mixed grasses and clovers, seed may be sown to produce pasturage for sheep and cattle. The radical leaves, which shoot up close to the ground, are continually cropped by stock, and afford excellent fodder.

MEDICAGO SATIVA

(*Lucerne*).

Dr. Stebler controverts the opinion that the common name of this plant is derived from the canton or town of Lucerne, in Switzerland, although he is at a loss to account for the appellation.

Lucerne was known to the ancient Greeks and Romans as a forage plant. It is mentioned by Virgil, and by several agricultural writers at the beginning of the Christian era. In Persia and other Eastern countries the plant is still used for horses. Throughout the American continent it is known by the Spanish name of Alfalfa

Wet summers diminish the culture of Lucerne in England, but dry seasons always restore it to favour, although English farmers have never appreciated its merits to the same extent as their Continental brethren. In part, no doubt, this is owing to climate, and in part to soil. Lucerne is essentially a plant for dry climates and dry soils. Cold, dry air has no ill effect upon it; but moisture, with or without heat, is directly prejudicial. Its duration depends more on the subsoil than the surface; indeed, the nature of the surface is of small moment so long as the subsoil is calcareous. In some parts of the Continent Lucerne remains as good a crop twenty-five years after sowing as in the third year of its existence, but in this country five years is the usual term, and seven years may be considered the fair limit of its vigorous life, even on suitable land. It is interesting to note the diverse surface soils on which Lucerne thrives, but investigation will always prove that whatever the nature of the surface may be there is a substantial agreement in the subsoils where this plant flourishes. The distance to which the tap-roots penetrate is almost incredible. In the first season they will often go down two feet or more. I have been told of one instance where a Lucerne root was traced unbroken to a depth of sixteen feet.

But for its unsightliness Lucerne would be more frequently sown for permanent pastures than it is, a practice advocated by the late Sir J. B. Lawes. The plant develops so rapidly that it stands far above the surrounding herbage, and imparts a broken and irregular appearance to the meadow. Still, it must be conceded that a certain proportion of Lucerne is desirable in every prescription of seeds intended for the formation of a permanent pasture where the subsoil contains abundance of lime.

In many localities Lucerne cannot be cultivated successfully, and to attempt to grow it on impervious clay, or on any cold adhesive land, will only be waging a fruitless war against Nature. Warm and calcareous soils are highly favourable, and a sheltered field, sloping to the south, will suit it to perfection, provided always that lime can be reached, for this the plant must have. Sometimes there is a strong desire to grow Lucerne on soil deficient in lime. To meet the case a heavy dressing of marl is applied six months before sowing; but the process is costly, and at best the effects are temporary, because it is impossible to bury the marl deep enough to feed the roots after they have pierced the surface soil. The richer the soil the earlier will Lucerne come to full development, and land should be chosen in which the roots will be able to strike down without undue resistance.

The principal point in the culture of Lucerne is to secure a thoroughly clean seed-bed. Weeds soon ruin the plant, and therefore farmyard or stable manure, although good in itself, is dangerous from the seeds it may contain. Ash of all kinds is beneficial, and, of course, artificial manure can be freely resorted to. Three hundredweight of superphosphate per acre is a good dressing. The best preparation is a crop of potatoes. As a rule, April is the time for sowing, and the seed must not be deeply buried. In England Lucerne is almost always sown alone, while on the Continent a thin seeding of corn is often put in with it.[1]

When well established the herbage must on no account be allowed to grow old before being cut; in fact, the plant should not be permitted to flower.

Lucerne is rarely made into hay, as the leaves are lost during the drying, and the process is exceedingly wasteful. The most convenient and profitable way of growing it is to sow a patch near the homestead, so that the daily portion

[1] My friend the late Mr. Clare Sewell Read told me that he had been most successful in obtaining a plant of Lucerne by sowing it in wheat, while he had never succeeded with it in barley, and seldom when sown alone. He attributed the failure among barley to the closer sowing of this corn as compared with wheat; and the failure when sown alone to the plant being smothered by annual weeds.

when cut has only to be carried a short distance to the stables. The plant is peculiarly rich in albumen, and is even more nutritious than Red Clover. When given alone, especially if cut very young, there is a possibility that cattle fed upon it may become blown, but when chaffed with good oat or barley straw Lucerne is a wholesome and valuable food. Several cuts are obtained in a year, making altogether an immense weight of keep. It is not worth while to sow seed unless the crop can be allowed to stand for at least three years.

Between 1893 and 1909 the area of Lucerne in England expanded from 17,617 acres to 64,908 acres, the extension amounting to 47,291 acres—a sufficient proof that the merits of Lucerne are increasingly appreciated, especially in seasons of drought.

ONOBRYCHIS SATIVA

(Sainfoin).

Sainfoin has been cultivated in this country for over two hundred years. It is essentially a food for sheep, and in pasturing the plant they do it no injury. Sainfoin is also useful for horses, but produces nothing like the quantity of green food that can be obtained from Lucerne.

In Norfolk and some other of the Eastern counties, Sainfoin takes the place of Red Clover, and is rarely allowed to remain down more than two years, generally only one. Against this practice nothing can be urged, for it is probably the best possible preparation for wheat. But the method adopted in Berkshire, Oxfordshire, Wiltshire, and Hampshire, of putting down Sainfoin alone for four or five years, has grave disadvantages. Gradually the Sainfoin plant diminishes, and every form of worthless vegetation increases, until the land becomes exceedingly foul. Instead of sowing Sainfoin alone, to remain down more than two years, the seed may with advantage form one of the constituents in a mixture of grasses and clovers for three or four years' ley. The sowing of Sainfoin alone is expensive, and more or less precarious. When combined with strong-growing grasses there is less risk of failure, and the grasses keep down weeds and prevent the growth of couch and other pests which almost invariably overrun a pure Sainfoin ley after the first two years.

In the green state Sainfoin is quite free from the danger of blowing cattle, and when made into hay it is an admirable and nutritious food. But the making of Sainfoin hay is rather a difficult business, particularly in catching weather. Like Trifolium, the plant has a hollow stem, which when cut is more quickly deteriorated by wet than any of the clovers.

There are two varieties, the common, and the giant or double cut, the latter being the earlier and more rapid-growing of the two, but decidedly less durable.

PETROSELINUM SATIVUM, OR CARUM PETROSELINUM

(*Sheep's Parsley*).

Parsley is frequently included in mixtures of grasses for uplands and sheep downs. Sheep manifest a great fondness for the herbage, which has been said to be a preventive of rot and red-water in that animal. Hares will visit gardens for the sake of the Parsley grown in them, and where this game is abundant it may be worth while to sow patches in the covers. The seed germinates so slowly that six or seven weeks may elapse before a plant is visible.

POTERIUM SANGUISORBA

(*Burnet*).

Burnet is a perennial, and a native of Britain. When sown alone the coarse unpalatable herbage is declined by all kinds of stock unless pressed by hunger, but it is eaten when associated with more tempting herbage. Seed is sometimes sown with grasses on poor chalky soils, where in dry summers sheep graze the closely cropped pastures. In mixtures of seeds for rabbit-warrens, Burnet may be employed with advantage, and it is occasionally used in limited proportion for pastures on light sandy soils.

Sowing Grass Seeds

The actual work of sowing grasses is simplicity itself, but as the germination of the seed and the equal distribution of the plant depend on the accuracy of the process, the details should be carried out with due regard to the serious loss which carelessness entails. The necessity for making the seed-bed fine and firm has already been explained. At sowing time the soil should be dry enough to allow the implements to work freely without any tendency to gather clods on the roller. Waiting for suitable weather may tax the patience severely, but it is worse than useless to enter into a conflict with Nature. All such conflicts inevitably end in the defeat of man.

The first operation is to run the harrow over the land to prepare it for the seed, which may be sown either by hand or by means of the seed-barrow. Some men are very skilful in spreading seeds equally, and on a still day their work may be successful. But grass seeds are light, and a gentle breeze is sufficient to make the sowing irregular. As the barrow delivers the seed near to the ground, it will, as a rule, distribute the grasses more evenly than the most practised sower by hand. But whichever method is adopted, there is a decided advantage in making two sowings. Half the quantity should be sown by passing up and down the land, and the other half by crossing the first sowing at right angles.

A bush-harrow is the best implement for covering the seeds, but a light iron harrow will answer, and the lighter it is the better. The object to be secured is that as many seeds as possible shall be thinly covered with soil. Grass seeds will germinate and become established when they are merely pressed upon the surface of the earth, provided they are not consumed by birds or scorched under a hot sun. But many seeds will fail to germinate at a greater depth than half an inch. Hence the necessity of a fine soil and shallow sowing.

The roller must promptly follow the harrow, and it makes a better and more certain finish to go over the ground twice in different directions with a roller of moderate weight than to accomplish the task at one stroke with a heavier implement. The importance of this part of the work will be made apparent if any spots are missed by the roller, for on those spots no grass will appear.

A good plant may often be obtained by rolling with the Cambridge or ring roller. Then sow the seed, and afterwards cross the land once or twice with the ordinary smooth roller.

Sowing grass seeds by the seed-drill is a modern practice, and for light sandy soils, especially in a dry spring, it has decided advantages. Under favourable conditions drilled seeds germinate freely, and endure summer drought when surface-sown seeds become malted. The covered drills afford protection from direct sunshine, the seeds have more certain access to moisture, and on corn that has grown too high in spring to render broadcasting a safe procedure the drill may often be available. In any case the coulters should be carefully adjusted for shallow sowing, and must be set as closely as possible. Sowing at two operations should also be regarded as a necessity. The harrow will not be requisite after drilling, but the importance of rolling firmly down is of even greater consequence than when seeds are got in by the hand or seed-barrow.

Hard and fast rules cannot be laid down as to the time for sowing. In a great measure it depends on the weather, and perhaps the easiest way of arriving at a conclusion will be to consider separately the relative merits of spring and autumn sowing.

Spring Sowing.—The term 'spring sowing' is here intended to apply to the months of March, April, and May, and as a rule it is better to sow early than late. In the early spring the soil is generally moist enough to keep the plants growing after they have made a start. But as the season advances the state of the ground becomes increasingly critical in this respect, and there is the risk that it may get too dry to maintain the grasses until the next rainfall. Then follows the extreme mortification of seeing a promising plant gradually wither away.

April is properly regarded as a safe and favourable month in which to sow, but if the seed-bed is ready and the land in working order by the middle of March, there need be no scruple as to putting in the seed. And should there happen to be signs of approaching rain, it is worth any reasonable effort to get the sowing done and the land rolled down before a change of weather can stop the work. It is an advantage to sow before rather than immediately after a shower, even supposing the land can be worked soon after rainfall, which may or may not be the case. The seeds gradually absorb moisture from soil and dew until rain comes, and then the plants rapidly spring up.

To sow later than the end of May is most hazardous. Even the middle of May will often be too late, especially on heavy soils, which in a dry season are also liable to crack, to the injury of the grasses.

Now comes the question as to sowing alone or upon a corn crop. This depends on the object in view, as well as on the length of the purse. Apart from the money consideration, each method has its advocates, and undoubtedly there are sound reasons to be urged in favour of either practice.

When a first-class pasture or an ornamental park sward is wanted as quickly as possible, and the cost is of no importance, the sowing of grass and clover seeds without corn will, in the majority of seasons, produce the most satisfactory pasture in the shortest time. Opportunity is thus afforded for early and frequent mowing, which tends to strengthen the young grass. A still more valuable service is rendered by the scythe in preventing weeds from seeding, and these are certain to be troublesome enough under the most favourable conditions. On the other hand, in an extremely dry summer, the shelter of a light corn crop will prove of great value to the young grass plant. It may even result in an actual gain of time. Without its aid there is a possibility that the grasses may stand still or completely fail for want of moisture. An established pasture when burned brown by the sun speedily recovers its verdure after rain, but young grasses cannot endure so fiery an ordeal.

The assistance rendered in checking weeds is another benefit derived from a corn crop. To appreciate its value, compare one of the finer grasses with some weed growing near, and it will at once be apparent that the delicate stripling has no chance of resisting its masterful neighbour. The annual weeds will die out by-and-by, but in the meantime almost every one of them will destroy some grass plants. It follows that the more abundant the weeds the thinner will be the pasture, and until the grasses tiller out and cover the ground the crop will be proportionately small. Unfortunately, both drought and weeds prove more injurious to the smaller than to the coarser grasses. The fact that corn assists in the battle against both foes is quite sufficient to account for the very general practice of sowing permanent grass seeds with a corn crop.

On the pecuniary value of the corn it is needless to dwell. The point is too important to escape attention, and, as a rule, this consideration is alone sufficient to determine the question in favour of corn. Of course a corn crop will levy the usual tax upon the land, and it should be clearly understood that the grasses are not to sustain the loss. A liberal top-dressing of cake-fed manure must be applied after the corn is cut, to compensate the grass for what the corn has absorbed.

One point is of utmost consequence if corn is not to injure the coming pasture, and this is the necessity of a very light seeding of corn. A heavy crop is harmful in itself, and involves further danger when it becomes laid. On the spots where a heavy crop is lodged the grass will probably be killed outright, and the slight additional gain derived from a full seeding of corn will be more than counterbalanced by losses in the grass plant, to say nothing of the labour of patching it afterwards.

The time for sowing grass seeds with spring corn will be either immediately after the corn is got in, or when it is only two or three inches high. It is well understood that the less forward the cereal, the better the chance for the grass.

On heavy, and especially on rich land, the choice of corn is open. It may be either barley, oats, or wheat, and wheat is always desirable for the grass.[1] For lighter soils barley or oats are available, and oats are preferable to barley.

Under certain conditions it answers well to cut the oats green, and turn the crop into hay or silage. This method of treating the herbage helps to keep down weeds quite as much as when the oats are allowed to mature, and it takes far less out of the land.[2]

[1] I have been most successful from an April sowing on a thin plant of wheat, and the late Mr. Clare Sewell Read said: 'I never find any difficulty in obtaining a plant from seeds, even in May, when sown with wheat, for then the ground is firm and the surface soil very fine. Often when the seeds fail in barley, the headlands round by the gates have a good plant, because there is fine mould on the surface and a solid bottom.'

[2] A well-known Scotch agriculturist states that he 'considers the best method of sowing to be with about two bushels of oats, to be cut green before there is any kernel. There is a large crop of useful fodder, the small seeds have beneficial protection while they require it, annual weeds are kept down, and the grasses get relief by the

Occasionally a field in autumn wheat is wanted for a permanent pasture, and there is no difficulty if the land be clean enough, and the grass seeds can be sown before the wheat is too high. In favourable weather the seeds may be put in even as early as the middle of February, as the corn will defend the young grass from injury by frost. Should the wheat be very backward, however, or stand thin on the ground, the sowing had better be deferred for a time. In the event of the land being at all foul, hand-hoeing must be resorted to, and this will open the ground for the grass seeds. The requisite harrowing and rolling will be beneficial to the wheat plant.

Sowing with Rape.—Notwithstanding all that has been said in favour of sowing Rape with grass seeds, I cannot recommend the practice. Instances can doubtless be cited where no injury has resulted. But the great objection remains that Rape necessitates feeding off the crop by sheep, and, when the Rape is ready, the grasses rarely have sufficient hold of the ground to bear grazing with impunity. The animals eat the hearts out of some plants, pull up many more, and altogether do a lot of harm to a young pasture.

Autumn Sowing.—Many writers have vexed their souls concerning the relative merits of spring and autumn sowing, without getting nearer to a solution of the problem. It cannot be solved at all by generalities, although the attempt has often been made. For practical ends it may be disposed of here by accepting necessity as a guide, and then perhaps there need be little or no controversy about it. The decision largely depends on the possibility of working heavy land in a wet spring. Sometimes autumn sowing is resorted to

early cutting at the stage most suitable for them to have full possession of the soil.' He adds: 'I have sown down one hundred and sixteen acres in this way. The same grasses, sown at the same time, and sometimes on parts of the same field, but with the oats allowed to ripen, have proved decidedly inferior.'

when a hot dry summer has scorched the life out of a spring plant. This is one of the many misfortunes to which the agriculturist is liable, but it does not touch the point now under consideration.

Were all the land of the United Kingdom light, probably the question would never have arisen. There would have been a general consensus of opinion in favour of spring sowing. It is the extreme difficulty of making heavy land ready for grass seeds before the spring is too far advanced which renders the state of the weather of so much more importance when sowing grasses than when sowing any other seed. Sometimes it is absolutely impossible to thoroughly pulverise a tenacious soil until May is far gone, and then it is very risky indeed to put in grass seeds. Thus late summer or so-called autumn sowing becomes imperative. Having reached this conclusion, it is satisfactory to remember that in addition to the chance it affords of making a thoroughly sound seed-bed, the temperature of the land in autumn is highly favourable to the germination of grass seeds, particularly in the North of England. Further benefit is to be derived from the clearing of successive crops of annual weeds, most of which would have appeared among the grasses of an earlier sowing.

The danger of an autumn sowing mainly concerns the clovers. Young grasses, especially of the stronger varieties, will stand much winter cold with impunity. Not so the clovers, although when established they also will endure severe weather unharmed. While young, a wet cold winter will almost certainly make an end of them. A retentive soil fosters a magnificent pasture containing plenty of clovers when once the plants are matured. Yet on such soils it frequently proves extremely difficult to obtain clovers from a sowing of seed in autumn.

As to the best month for autumn sowing, it must not be forgotten that grass seeds are peculiarly liable to be 'malted' under a burning sun when the ground is not moist enough to keep the plants going. A shower on a dry soil, followed by scorching sunshine, is quite sufficient to destroy the germs, and many failures of autumn- as well as of spring-sown seeds are attributable to this cause alone. Even when the failure is not total, it is a serious matter if the finer grasses are killed and only the coarser varieties survive. After the stronger varieties have had a six months' start it is very difficult to establish the finer grasses. As a consequence the pasture may always be inferior in quality.

Supposing, however, that the soil is not too dry, and is otherwise in good condition, sowing in August is to be commended, and the earlier the better. The plants will then have a chance of becoming fairly strong before winter sets in. The first week in September is, in most seasons, the latest date at which a permanent pasture can be sown down with any reasonable hope of a satisfactory result, except in the extreme South.

I have formed a good grass layer by sowing in autumn with winter oats, and the grass came so forward as to be fit to carry stock in September of the succeeding year.

Every field sown in August or September should undergo a searching examination in the following spring. If it is evident that the clovers and finer grasses have failed, more seed of the missing sorts should be sown before the grasses become too strong. There need be no anxiety as to the effects of spring frosts, for the grasses will afford the young clovers friendly and effectual shelter.

The Immediate After-Management of New Pastures

A SPRING sowing of grasses is made at a time when atmospheric changes are sudden and severe, and grass seeds are not so well constituted for resisting these violent changes as corn and other heavy seeds. Besides, the spring is never so dry and cold as to prevent the growth of weeds, nor is the May sun hot enough to kill them; but after sowing, a long spell of unfavourable weather will seriously retard the grasses. Meanwhile the ground may be covered with chickweed, groundsel,[1] and other weeds. As these extend, the chances of the grasses diminish, until at length it is possible that only a few spots will be found on which they show sufficiently to prove that there would have been a crop had circumstances been favourable. In a backward spring early sowing increases the danger of the grasses being smothered by weeds.

But if early sowing has its perils, late sowing is not free from them. From the former arises the possibility that the young grasses will be injured by weeds, and from the latter that before the grasses are sufficiently established to endure great heat, they may be scorched beyond recovery by fierce sunshine. Or the soil may be so dry that the germination of the seed is dependent on rain, and if only a brief shower falls, the seeds will start. Then, should there come the dry blast of an east wind, or burning drought, every seedling will perish.

[1] Groundsel will actually flower when the thermometer stands near the freezing point. Humboldt observed the plant growing in the upper reaches of the Andes, just below the region of eternal snow, where the sun has little power and where hurricanes are incessant.

I am not conjuring up difficulties for the sake of saying 'There is a lion in the way,' but rather to show that for so important and costly an undertaking as laying down land to grass there is absolute necessity for insisting on the cleanest possible seed-bed. Even when the farmer has done his utmost to clean the land, plenty of weeds will spring up. It is then a question of precedence. If the grasses come quickly, the annual weeds do little harm when promptly checked, but if the weeds obtain a strong lead, the injury to the grasses may prove serious, perhaps entirely destructive. Happily, the majority of seasons favour the sower, but that does not relieve him from the obligation of taking every reasonable precaution to ensure success under what may prove to be very adverse influences. Possibilities must be considered, and by being forearmed the probability of success will be enhanced.

In the immediate after-management of newly-sown grass the advantages of dispensing with a corn crop can be fully realised. It is impossible to render active assistance to grasses sown in corn until the crop is cut and carried; but when grass is sown alone the plant can be mown with a sharp scythe as soon as it is a few inches high, and the benefit will speedily be visible. After mowing, the roller should be put over the land again, which will help still further to consolidate it, and to give the young plants a firm grip of the soil. The more frequently the pasture is mown and rolled during summer, the more rapidly will the ground be clothed with verdure. By August or September, in a favourable season, the grass may be allowed to grow on to produce a small lattermath hay crop, after which it will carry horned stock, eating cake, through the autumn. The young crop should never be allowed to grow unchecked until winter sets in, as the herbage has a tendency to rust and rot away.

Although constant mowing will get rid of groundsel and other annual weeds, it is powerless against such pests as docks and coltsfoot. These can only be removed by a spud or narrow hand-hoe, and for the sake of the future of the pasture it is well worth while to give these strong-growing weeds early and persistent attention. Thistles can be eradicated by repeatedly chopping them off near to the ground with an old scythe until they bleed to death.

Should it be found by the middle of May that the grass has failed, no time must be lost in shallow-ploughing or cultivating the land, breaking it down again into fine tilth, and re-sowing. Supposing, however, that the seed has taken over the main area, leaving some bare spots, these must have their crust broken with a hoe, followed by a heavy rake, and be re-sown, raked over, and again rolled down. When the plots which have missed plant are too large for hand work, the harrow can be employed instead of the hoe and rake. It will never do to assume that the crop is all right. Rather assume that it is not perfect until examination has shown the contrary.

Grasses and clovers sown with corn should not be allowed to suffer injury from thistles and other coarse weeds. The destruction of these pests is necessary for the corn, and is imperative for the success of the grasses. Immediately the corn is off the ground the whole field should be carefully examined to ascertain whether there is a plant or not. After a wet harvest there will be no difficulty in deciding promptly. But in a dry season I do not advocate a hasty judgment. The dry stubble, the parched ground, and the brown herbage all help to mislead. A few hours of warm soaking rain may put a new complexion on an old face in so brief a time as to produce an almost magical effect, and therefore it is not wise to be in too great a hurry to pronounce upon success or failure. More than one instance has come to my knowledge where a spring sowing of permanent grasses has been followed by a dry summer, and on cutting the corn scarcely any grass was visible. A top-dressing of four hundredweight of bones and superphosphate per acre was applied with such good effect that in the following summer a heavy crop of hay was cut.

If the plant is evidently all right, there can be no doubt that it will abundantly pay to give a top-dressing of farmyard manure, or some good artificial, to help the young grass into vigorous growth. Bare spots caused by the laying of the corn or from any other agency must be lightly broken, sown, and rolled down again. It will be quite necessary to look these patches over in the following spring to see that they have passed safely through the winter, otherwise they must be sown once more.

Should the failure be total, it will generally be impossible to smash a hard stubble, and get it clean, fine, and firm by the first or second week of September; and therefore it is usual to defer re-sowing until the following spring. On two or three occasions I have risked sowing grass seeds on an unbroken stubble, after the manner common with Trifolium. In each case the stubble was unusually clean, and directly the corn was carried, a heavy drag was put over the land and the seed was bushed in. The success was very marked indeed, but I do not feel justified in drawing large inferences from a few experiments of this kind.

A pasture sown with corn will not, after the corn is cut, be in the same condition for grazing as when grasses are sown alone. In the latter case the care and attention that can be devoted to the plant through the summer make all the difference. After a showery season horned stock can sometimes be turned on to a stubble containing young grass without inflicting serious injury. But if there are occasional instances of this kind there can be no doubt about the folly of permitting sheep to graze it. Sheep bite extremely close, and with a snatching movement which uproots an immense number of young plants that have not sufficient hold to bear the strain. Another fact is worth consideration. Both cattle and sheep, if allowed to graze too soon, are apt to pick out certain grasses and clovers for which they have a partiality, leaving others to seed or to develop into ugly tufts.

Whenever this happens it is wise to let a man go over the ground two or three times, and cut these tufts down. The new growth will afterwards be eaten close.

It is unquestionably sound practice to mow a permanent pasture the first year after sowing, and on some soils it may be advisable to mow in the second year also. The loss to the land must, of course, be restored, either by a liberal dressing of farmyard manure, in autumn, or by artificials rich in phosphates and potash, applied in spring.

In the early management of autumn-sown grasses, the object to be kept constantly in view is the promotion of free growth before winter sets in. Topping the young grass with the scythe and rolling will prove advantageous to the plants in helping them to cover the ground and become firmly rooted. Immediately the growth begins in spring mow once more, and a final rolling is also essential. After an autumn sowing it is especially necessary to cut the hay crop very early. When it is carried, cattle may be turned in to graze.

Several of the finer grasses, if permitted to seed while young, are so weakened that they die, and on some soils they appear to perish more readily than on others. This does not show that such grasses should be excluded from a prescription for a permanent pasture, as some writers affirm. It would be just as reasonable to say that because certain varieties which revel in a dry soil disappear after a succession of wet summers, therefore they ought to be omitted. A pasture is laid down that it may yield nutritious herbage, not that seed may be saved from it. Grasses which require three or four years to attain maturity—and there are varieties which do not reach their highest vigour in less time—must of necessity be weakened or destroyed by producing seed in the first or second year after sowing, just as animals are permanently stunted by allowing them to reproduce their species at too early an age.

The opinion is widely entertained that the critical period of a pasture is the third or fourth year after it has been sown. But if a pasture begins to fail about that time, it is probably attributable to mismanagement and starvation. No farmer supposes for a moment that he can for several years in succession take heavy crops off arable land and put nothing on it. Yet this is a very common delusion concerning grass land. And I say most emphatically, that the man who thinks it reasonable to treat either a new or an old pasture on that principle deserves to find it deteriorate in quantity and in quality also. Liberties of this kind are sometimes taken with a rich old pasture, and the injury may not at once be apparent; but it is most unreasonable to expect that a young pasture will become established under the starving system and at the same time yield heavy crops.

One cause of the early deterioration of some new pastures is no doubt traceable to grave faults in the prescription of the grasses sown. Too many farmers are content if they can only see 'something green,' without bestowing a thought as to whether the 'something' is good or bad. So long as men will only pay about half the value of a first-class prescription of permanent grasses and clovers, dealers will be prepared to supply so-called permanent mixtures, consisting mainly of annual varieties of Rye Grass, Yorkshire Fog, Tussock Grass, and other cheap seeds utterly unsuitable for the purpose.

In a subsequent chapter reference is made to feeding an old pasture by supplying the animals upon it with cake, and there is no better means of enriching the land. But if the plant cannot be safely fed off until about eighteen months after sowing, obviously some other means of stimulating the pasture must be adopted, and this is why I strongly advise a top-dressing of farmyard manure after corn is carried, or as an alternative an application of artificial manure in spring.

The Rothamsted and my own experiments at Kidmore have demonstrated a fact which I am anxious to emphasise.

After every care has been exercised in selecting suitable grasses and clovers, and a plant has been established, the herbage of any piece of grass will eventually depend on the after-management. If a field which has been judiciously sown be divided into several portions, and each portion is subjected to distinct and continuous treatment for successive years, a decided difference in the herbage of the several parts will become manifest. Certain manures encourage the growth of certain grasses, and indirectly effect the destruction of those species which are not benefited, by enabling stronger neighbours to choke them. Those who have carefully observed the results obtained in the late Sir J. B. Lawes's experimental grass plots will adequately realise the importance of applying suitable manures, not merely for the purpose of augmenting the crop, but as a means of maintaining or destroying some of the grasses.

Manuring is not the only way of effecting changes in the character of a plant of grass. Some varieties are specially adapted for grazing, others for making into hay. A fine old pasture which has been fed for many years will often yield a miserable crop of hay, and may be utterly ruined by being mown for several consecutive seasons. Conversely, a meadow which has been mown for years and kept in condition by annual top-dressings may prove altogether unsatisfactory as a pasture. Grazing gives all varieties of grasses, except a few which will not bear treading, a full chance of existence; while haying fosters the growth of those grasses which come to maturity at a particular period of the year. Some of the most valuable pasture grasses are often entirely absent from good old meadow land. It is therefore desirable as far as possible to reserve meadow land exclusively for mowing, and pasture land for grazing.

THE MANAGEMENT AND IMPROVEMENT OF OLD GRASS LAND

IN 1872 there appeared in the Royal Agricultural Society's Journal an account by Mr. H. S. Meysey Thompson of a personal inspection made by him of the grass lands of England and Ireland. His laborious journey included visits to the most famous grazing districts, and it was undertaken when farmers were at the height of their prosperity. The following are his conclusions :—

'1st. That although very excellent management is to be met with in parts of our best grazing districts in Leicestershire, Northamptonshire, Gloucestershire, Somersetshire, and several other counties, this must be considered quite exceptional, and the treatment of the bulk of the grass land of the country is very unsatisfactory.

'2nd. That our grass lands, if properly managed, would be easily able to meet the demand made upon them for an increased production of meat, even if the supply required were greatly in excess of the present rate of consumption.

'3rd. That money judiciously laid out in improving grass land makes a better return than money laid out on arable land.'

These remarkable statements are as worthy of profound attention from those who possess old pastures, as from those who contemplate laying away arable land to grass. Unfortunately, much of the grass land of this country is now in very little better condition than at the time Mr. Thompson wrote. There is not the least exaggeration in saying that thousands of low-lying meadows and upland pastures are not yielding half the produce which could be obtained from them were the land in better heart. These pastures grow inferior hay and little of it, the production of milk is restricted, and the capacity for fattening stock is diminished. The result is an enormous national loss, and the truth must be told that this loss is almost entirely avoidable. The prime cause is negligence, begotten of the mistaken notion that a pasture is self-supporting. There is a very general assumption that the owners and occupiers of grass lands are not only relieved from the anxiety and expense of arable tillage, but that they are under little or no obligation to make any return to the soil for all that may be taken from it in the form of hay, milk, or meat. I propose to specify some of the influences which have reduced many English pastures to their present unsatisfactory state, and to suggest means of restoring them to fertility.

One of the principal causes is the practice of taking hay crops for several successive years without giving any adequate return in manure. The necessity for treating arable land liberally is never disputed for a moment. Yet the arable land has the advantage of being constantly broken up and enriched by rain, air, and other of Nature's fertilising agencies; while a pasture is, by its fixed condition, debarred from the benefit of all cultural operations, except the use of the harrow and roller. If properly drained, grass land can generally be maintained in the full tide of fertility by judicious manuring alone; but this is often negligently or wilfully withheld.

The prevalent idea that continuous haymaking is inimical to the welfare of a meadow is entirely illusory, and probably arises from the failure to give any return to the land for the crops taken from it. Of course the hayrick can be made the instrument of impoverishing grass land more quickly than overstocking; still, it is not so much the cutting of a meadow which is injurious, but withholding the top-dressing of manure or compost which all mown grass should annually receive.

Again, immense damage is done to some meadows by cutting the crops foı hay very late in the season. Many of the grasses have time to form and ripen their seeds, and nothing exhausts plants so much. Some of the finer grasses cannot safely be taxed in this way. They may endure the ordeal once or twice, but if the drain on their resources is frequently repeated they gradually dwindle away. So long as these grasses are mown early, or grazed, they are perfectly perennial, as is conclusively proved by their continued existence in some of the finest old pastures of the kingdom where they have never been allowed to seed. To manage any pasture or meadow in such a way as to exterminate some of its most valuable and nutritious grasses is surely killing the goose that lays the golden egg. And this is exactly what is done by constant greed for the rick. Of course an early crop means a smaller bulk of hay, but unless the turf is exceptional in character the quality is higher than from a later cutting,[1] and

[1] In the *Journal of the Bath and West and Southern Counties Society* for 1890–91, page 375, it was suggested by the editor that I should undertake a series of experiments with the object of determining 'the value of aftermath, according to the period of the first cutting; also to determine the relative food-producing powers of a meadow when various times of cutting were resorted to.' I gladly assented to the suggestion, and my friend Dr. Voelcker was kind enough to undertake the chemical part of the experiment. The results were published in the Society's Journal, Vol. II., Fourth Series, from which I have extracted the following comparative table of five plots in the same field, having the same aspect, soil, and other conditions calculated to ensure uniformity of growth:—

Analyses of Grass from the Plots—First Cutting. In Dried (at 212° F.) State.

Plot	Date of Cutting	Soluble Albuminoids	Insoluble Albuminoids	Digestible Fibre	Indigestible Fibre	Soluble Carbo-hydrates, &c.	Mineral Matter (excluding Silica)	Silica
No. 1	June 3	3·41	7·82	31·12	26·67	23·43	5·75	1·80
„ 2	„ 17	2·28	6·67	32·61	30·87	20·93	5·02	1·62
„ 3	July 2	1·46	6·92	32·14	31·70	20·41	5·51	1·86
„ 4	„ 15	1·81	6·98	27·35	31·10	24·78	5·69	2·29
„ 5	„ 29	3·44	5·67	28·88	30·47	24·33	5·10	2·11

the loss of bulk in the first instance results in a net gain, for the aftermath will be the greater, and the pasture will not be damaged for future seasons. Where grasses perish, the gaps are almost certain to be filled by worthless or noxious forms of vegetation, and thus the herbage diminishes in value so long as a false system of management is pursued.

Another source of injury to pastures arises from the manner in which grazing is conducted. It should not be necessary to repeat so trite a remark as that land is never enriched by the droppings of cattle fed exclusively upon its herbage, but, on the contrary, must by degrees become the poorer for supporting the lives and increasing the weight of the animals which graze upon it. In milk and flesh the land yields its produce in highly concentrated forms, and without external aid the process of exhaustion must of necessity go on. But when the herbage consumed is supplemented with cake, corn, roots, hay, or other extraneous food, benefit is conferred on the pasture in addition to the advantage which the animals derive from it. The improvement will, of course, be gradual, and its progress be regulated by the quantity and the quality of the additional food supplied. In this extra feeding of grazing animals there is a simple and economical means of enriching a poor pasture, and the increased weight of the stock is an immediate if only a partial return for the outlay. Agricultural chemists tell us, and their analyses are supported by experience, that animals only assimilate one-tenth of the nutritious qualities of cake or other highly concentrated feeding stuffs, and that the remaining nine-tenths, after passing through the cattle, are available for vegetation, in a form specially adapted to meet the requirements of plant life. This explains the marked improvement which is always observable when grass is depastured by cake-fed cattle—an improvement superior to that effected by a dressing of farmyard manure, because none of the valuable elements are lost by fermentation. And this fact suggests the economical aspect of the practice. The carting of heavy bulks of manure is avoided, and the land at once has the benefit of the droppings. When manure is stacked in heaps, or is allowed to lie in the farmyard, some of its most fertilising constituents drain away or are dissipated in the atmosphere. It will also be evident that to graze a pasture by day and fold on the arable at night is a very ingenious device for ruining grass land. Even when sheep are helped with cake, it is no sufficient compensation for their absence during twelve out of the twenty-four hours.

A further means of deteriorating grass land is the practice of allowing pastures reserved especially for horned cattle to be overstocked. When an ox-pasture is eaten down so bare as to allow the roots of the more succulent grasses to become scorched, it is a serious injury, not only for that year's feed, but for subsequent seasons. In one instance, during a hot summer I hoped by a liberal allowance of cake to make a pasture carry more stock than the crop justified, and the result was disastrous to the plant. On the other hand, an established sheep-pasture can seldom be cropped too closely for maintaining constant growth of the sweet fine herbage of which it should consist.

Widespread indifference prevails as to the predominance of such weeds as cowslips, primroses, orchids, daisies, and plantains, although these plants frequently show that the soil is in such a condition as to be incapable of maintaining nourishing herbage. The mere presence of these weeds and of barley and brome grasses is an evil in itself, and they indicate that the land is starved, just as tussock grass, rushes, and sedges prove the need of drainage. Thistles, docks, coltsfoot, and other large weeds may also abound, and they cannot be eradicated without the constant use of the scythe and spud. In a foul pasture the weeds are generally so mixed up with what good herbage there may be, that they can only be improved out of existence as better grasses are induced to take their places. A heavy dressing of salt applied after weeds have been cut will kill a large proportion of them, and an application of gas-lime has been known to effect a surprising change in the herbage of an inferior pasture. The folding of sheep thickly will also produce marked benefit on poor upland grass if the animals are at the same time fed with corn or cake. They should be penned on the ground long enough to make it as brown as a fallow, and then many weeds will be killed outright. This practice differs very much in its effects from that of giving sheep the run of the land. Whatever discourages the growth of rough herbage encourages that which is better. It is equally true that, however good a pasture may be, it has only to be treated with a policy of masterly inactivity, and in time it will revert to the waste condition of a moorland.

A succession of wet summers is another fruitful source of injury to pastures. The bulk of herbage forced from them during warm damp seasons tends greatly to their impoverishment, and some of the grasses which are more especially adapted for dry soils will probably perish. Well-drained land naturally suffers least. Land not well drained becomes sour and unwholesome, and only the sedges and coarse water-grasses survive.

Hitherto nothing has been said about seed, and it may be frankly admitted that with liberal management it is quite possible to restore the fertility of some pastures without sowing seed. But the remedy will take time, perhaps many years; and it is a penny-wise and pound-foolish procedure to occupy a long period in making an improvement which might be effected in a single season. The outlay beyond that necessarily incurred in carrying out the improvements already suggested is very trifling. In every case where the plant stands thin on the ground it will pay to sow a few pounds of the finer grasses and clovers per acre. A farmer I am acquainted with sows every autumn on an old pasture twenty pounds of grass seeds per acre, because he has found by experience that the result is an increase of a ton of hay per acre in the cut of the following year. The seed may be sown either before the grass starts growth in February, or late in August or the beginning of September. On damp land preparation should be made by an application of salt to the most weedy parts, and a severe dragging over the entire surface. A well-mixed compost of lime, the contents of ditches, and any other available rich material, should be distributed over the whole meadow, and the seeds can be sown on any day when the ground is dry enough to permit the roller to be used. To ensure germination it is important that the bush or chain harrow should precede the roller. The meadow should then be laid in for hay, and after the crop is cut cattle may be allowed to depasture the land, but sheep must not be admitted until the following year. Upland pastures may be treated in a similar manner.

For destroying moss there is no better dressing than two cart-loads of lime mixed with eight cart-loads of light loam per acre. The heap should be turned several times until the lime is thoroughly slaked and well incorporated with the loam. After dragging the turf with heavy iron harrows, the compost should be evenly spread. There will soon be a marked improvement, and a full return for the outlay.

The effect of dragging a pasture is not everywhere appreciated at its full value. The mechanical action breaks up the congested surface, allows the atmosphere to penetrate to the roots, and thus promotes a free and healthy growth of the plants. It also enables the grasses to absorb and derive benefit from any fertilising agent or compost which may be applied to the surface, instead of allowing part of the dressing to be washed away by the first heavy rain.

Breaking up Old Grass Land

Some pastures are so unproductive and foul with weeds, that in order to secure better crops the only satisfactory course is to break them up, grow turnips or other roots for a year or two, and then sow good permanent grasses and clovers. This is always a serious proceeding; but if, in addition to breaking up the land, there be also the willingness to incur the cost of paring and burning, I believe the operation may often be worth undertaking, provided the soil is suitable.

A poor mountain pasture, however difficult it may be to ameliorate its condition in other ways, must on no account be destroyed. Paring and burning will almost certainly fail to increase its productiveness, and it is well-nigh impossible to form a sod on such land by artificial means. Nor will burning answer with any light sandy soil. Unless clay is a principal constituent, burning is to be avoided, for it will only do harm. As a rule, peats and thin clays resting on chalk are benefited by the process. The colour of the soil as a rule indicates whether or not burning will be advantageous. A bronzy black soil is fair evidence of the presence of protoxide of iron. Where this chemical compound exists in quantity burning will generally improve the soil, for the protoxide is injurious to vegetation; by the action of fire it becomes a peroxide, which is beneficial to plant life.

An old and easy mode of discovering whether land will benefit by burning is to place sods in a large iron pot or kettle, closely covered, and place over a gentle fire. The heat must not be sufficient to create a flame. Gradually the sods will char away to ashes. If the land is suitable, the ashes will be red and powdery, mixed with a few black particles, and when put into water will make it more or less muddy. In the proportion that the water holds the ashes in suspension will the land be benefited by burning. If the land is unsuitable, the ashes will be sandy, and instead of making the water thick they will be precipitated to the bottom, leaving the water almost clear. Supposing this experiment to be in favour of the operation, even then only a small area should be tried until there is conclusive evidence that the proceeding would be advantageous.

There is no necessity to pare the soil deeper than three or four inches, and instead of the spade or turfing iron, a paring plough made for the purpose is to be preferred, especially as it leaves the slice of soil on its edge, so that a drying wind soon fits it for being gathered into heaps. To economise labour, many small fires scattered over the field have been advocated, and undoubtedly they save much carrying of the turf and facilitate the spreading of the ashes. But small heaps are very wasteful. It is almost impossible to prevent them from flaring, and that is ruinous. Large dense masses, however, can be burned slowly and evenly, and at a comparatively low temperature. This point is worth attention, for it makes an enormous difference in the fertilising value of the ash. The inorganic constituents of the soil are rendered soluble when burned slowly, and become more insoluble when overburned.

The effect of burning is to get rid of all the organic matter. But the mineral constituents—with the exception of nitrogen—remain, and they are so transformed by fire as to be easily assimilable by future crops.

The consumption of all the organic matter by fire is of course a destructive process, and in itself involves a considerable loss, but the effect of fire upon the inorganic substances goes far to neutralise this loss. The soil is rendered capable of assimilating ammonia, nitrogen, and other plant foods more rapidly than before. Fire reduces clay to a friable, disintegrated condition, when the staple readily absorbs fertilising matter. The late Dr. Voelcker conclusively proved this to be the case, and that the effect of heat is to transform some of the mineral elements from an insoluble to a soluble state. For instance, as the soil probably contains various compounds of silicates and of lime, the lime becomes liberated by the heat, and, under the high temperature, attacks the silicates, sets free part of the potash from its insoluble compound, and converts it into soluble plant food. This liberation of potash has probably more to do with the success of burning than any other result consequent upon it. Thus the action of fire effects chemical and physical changes in the soil which are of material advantage to the plant life which follows. Burning has really very much the same effect, but in a more intense degree, that liming has in sweetening the soil, and in setting free quantities of inorganic matter which were previously in a locked-up condition.[1] An admixture of lime with the ashes greatly augments their value for clay soils.

Burning also destroys the noxious forms of vegetation, and the land is freed from bots and grubs, and other destructive larvæ of insects.

It is usual to take at least one crop of roots immediately after burning, and in such a case a mis-plant is rarely known. The ashes absorb so much moisture from the atmosphere, and give it up so slowly, that turnip seeds have plenty of time to germinate, however great the heat. Even the fly is seldom troublesome on newly-burned land.

[1] The following is Sir H. Davy's analysis of the ash of burnt turf:—

	Parts
Carbonate of lime	80
Sulphate of lime, or gypsum	11
Charcoal	9
Saline matter, principally sulphate of potash and muriate of magnesia	3
Oxide of iron	15
Insoluble earthy matter	82
	200

Temporary Pastures

Although a large proportion of the cultivated soil of the United Kingdom is perfectly adapted for being laid away to pasture, it is unquestionably true that certain soils do not take kindly to permanent grasses. There are also cases where absence of fences and want of funds to make them, the cultural preparations and expense of the seeding, combine to render the creation of a permanent pasture impracticable. In relation to this subject I published a short paper some years ago which had this question for its title: 'Is there no Alternative?' As an answer I ventured strongly to press upon the attention of agriculturists the necessity of adopting the system of alternating grass with corn and roots as a means of enabling them to work with less capital and of reducing the labour bill by at least one-third. The late Mr. Clare Sewell Read gave the sanction of his high authority to this practice by publicly stating that he considered it to be the only possible way of meeting modern conditions of agriculture.

Several of the reasons why some land will not grow a satisfactory permanent pasture are very ably stated by one of the greatest French agricultural authorities, Monsieur H. Joulie. In his essay on 'Permanent and Temporary Meadows and Pastures,' for which the Société des Agriculteurs de France awarded him a gold medal, he says:—

> 'At first the grass plants find a soil suitably dressed with farmyard or other manure, that is to say, containing all the

elements necessary for their growth. So they grow vigorously. But little by little the soil becomes more compact, the subsoil more dense, and the rain or water of irrigation penetrates with greater difficulty. During the droughts of summer the moisture rises up less easily from the subsoil, and thus, from physical causes, the production settles down to a normal level. In time the chemical condition of the land also undergoes a material change: not only is the layer of soil which is occupied by the roots rendered incapable of supplying a sufficiently large amount of the elements necessary to the vegetation, but, owing to the continued accumulation of vegetable débris, the layer of soil in which the roots live at length becomes sour, even where the earth may originally have been calcareous, and may still be so in the underlying layers, so that the good plants tend to disappear and give place to a vegetation which is characteristic of sour land.'

After giving the reasons and experiments which prove his case, Monsieur Joulie adds:—

'From all that has been stated we can now draw the following practical and economical conclusions:—

'1st. That the cultivation of roots and cereals deprives the soil of nitrogen, whilst that of grass and leguminous plants, temporary or permanent, on the contrary, causes it to accumulate in the soil. That nitrogen being the most expensive manure to buy, it is not economical to devote part of the land permanently to arable and part to grass, for while the one uses up the nitrogen, the other accumulates it in excess. On the contrary, it is preferable to alternate on the same piece of land the cultivation of roots and cereals with that of grass leys, so as in a measure to repair by the second the loss of nitrogen which the first cause to the soil. By this means cultivation can be kept up indefinitely without purchased nitrogen, provided that the land be maintained in a fit state of richness as regards the mineral elements which are indispensable to healthy vegetation.

'2nd. The practical application of this principle is, that the temporary occupation of the land by a grass ley for two or three years, which takes its turn in the rotation of crops, should be preferred. We thus secure the improvement of the soil obtainable from the cultivation of Leguminosæ (clover, lucerne, vetches, &c.). But as this class of plant will not succeed on every soil, temporary "leys" with gramineous (grass) herbage ought to give, where leguminous plants do not succeed, analogous, if not equally good, results, and so assist materially in solving the problem of producing cereal, root, and other crops with increasing economy.'

Upon nearly all soils alternate husbandry may be adopted with immense advantage, and on light sandy soil, where a satisfactory pasture is rarely formed, it is far wiser to sow a temporary than a permanent mixture. The periodical breaking-up of the land at the end of every three or four years, and its treatment as arable for one or two seasons, will render it capable of again yielding valuable crops of grass.

The heavy crops that can be obtained from artificial grasses during a limited number of years are, no doubt, partly attributable to the judicious selection of grasses and clovers. But other influences are at work. The continual use of manure has put the land into good heart, and cultural operations have allowed the atmosphere to set free the elements which grasses readily assimilate. There are, unfortunately, plenty of instances where improper seeding or starved land renders a profitable permanent pasture impossible. To these preventable causes I am not now alluding, but to soils which, in despite of fair treatment, agglomerate and become impervious to atmospheric influences, and refuse to give up the necessary elements for the continuous free growth of grasses.

Apart, however, from this question of an unsuitable soil, there are weighty reasons for the adoption of a system of alternate husbandry. Two-thirds of many farms might with advantage always be in artificial grass. A great saving is effected in tillage operations, horseflesh, and labour, and the land breaks up at the end of the term in excellent condition, full of clover roots as a store of nourishment for the succeeding grain crop. The custom of maintaining agricultural holdings that are almost entirely arable, or almost entirely pastoral, has failed to meet the necessities of our time. What is wanted now is a combination of arable and pastoral husbandry, so that when corn does not pay and stock is profitable, or *vice versâ*, each occupier may obtain benefit from one branch of his business. The grazier would be profited in being able to winter his own stock instead of selling it to make a winter's manure for the arable farmer. On the other hand, the arable farmer would not then, as now, be compelled to sell his stock immediately his roots were exhausted, or pay the grazier to summer the animals for him. When neither arable nor pastoral farms yield a profit, the system I am advocating has the merit of reducing expenses to a minimum.

The specialising of agriculture has been carried to injurious excess. Great arable farms, without enough pasture to keep half a dozen cows, and large grazing farms that are wanting in sufficient arable to grow straw and roots for winter consumption, should both be regarded as evils. The admirable system, pursued in Lancashire and in Scotland, of annually laying away in artificial grasses a proportion of each farm for a period of three or four years, is so successful that it is surprising the practice has not long since been adopted all over the country.

On midland and southern farms the custom long prevailed of sowing Broad Clover alone, or in occasional instances with an admixture of Rye Grass. The more profitable method is, however, gradually advancing in favour, and every year an increasing number of farmers find that the best results are obtained from clovers combined with such heavy-cropping grasses as Cocksfoot, Foxtail, Timothy, and Rye Grass. I am fully persuaded that the general adoption of short-term leys will prove to be a substantial gain. In itself the system of temporary pastures is good, and a means of good, for it opens up a superior method of farming which can be conducted with a smaller capital than is necessary for the management of a purely arable farm.

The assumption that there is no alternative between the old four-course system and laying down land to permanent pasture will not bear a moment's examination. The result of this fallacy is fraught with mischief, and entails an enormous loss on the farmers of this country every year. The alternate system offers a remedy of proved value, the adoption of which will tend very materially to turn a deficit into a favourable balance. Cocksfoot, Timothy, Italian Rye Grass, and other strong-growing grasses, not only produce heavy crops of nutritious hay, but they smother weeds and keep the land clean; that is, supposing it to have been in a reasonably clean condition when sown. The alternate system will justify neither slovenly preparation nor foul seeding. Those who sow rubbish will assuredly reap as they sow. The seeds sold for leys sometimes consist largely of Goose Grass and Yorkshire Fog, with a liberal sprinkling of sorrel and docks. Let no man who puts such a vile mixture into his land blame the system I am advocating for the miserable results he may obtain.

One Year's Ley.—For this purpose the varieties must obviously be restricted to those which yield a large and immediate return. Either Annual or Italian Rye Grass will form the basis of the mixture, and an addition of Perennial Rye Grass will generally be desirable, as also a very small quantity of Cocksfoot and of Timothy. For grazing, Broad Clover is preferable to Cow Grass, and White Clover will also be a necessary constituent. But for hay, Alsike should replace White Clover. Where Trefoil is indigenous, especially if it appears freely, seed of this clover is not wanted in a mixture. In other cases a proportion of it may be an advantage to the crop.

Two Years' Ley.—The extended duration of the ley justifies an increase of Timothy and Cocksfoot, but it is very easy to overdo the latter. Italian will almost certainly be preferable to Annual Rye Grass, although the proportion must be diminished, and a larger quantity of Perennial Rye Grass be substituted. Alsike and White Clover should be more freely sown, and the weight of Broad Clover be slightly reduced. A heavier total seeding will be necessary to make a satisfactory two years' ley than is required for a single season, and more regard must be paid to peculiarities of soil and to the object in view.

Three Years' Ley.—Besides retaining Timothy, Alsike, White Clover, and Perennial Rye Grass, it is still an advantage to include some Italian Rye Grass, Broad Clover, and Trefoil for the sake of the first year's produce. The value of the hay will be increased, and the pasture will be more palatable to stock, by partially replacing Cocksfoot with Meadow Fescue; Foxtail may also be introduced. When the ley is needed for pasturage, an addition of Hard Fescue will render good service by making the bottom grass dense, and the Broad Clover can then be supplemented with Cow Grass. On chalky soils either Sainfoin or Lucerne may be desirable. The total quantity of seed will approximate more nearly to that required for a permanent pasture, although the finer varieties will be omitted.

Four, Six, or Eight Years' Temporary Pasture.—There is a general impression that, for so long a term as four years, a permanent prescription should be employed, but this involves needless expense. Some of the finer grasses that are properly included for permanent pastures would be wasted if sown for four years only; they would scarcely become established until the term had nearly expired, and would certainly yield no adequate return for the outlay incurred. For the longer periods of six or eight years some modifications in the prescription must be made, but these will depend on the circumstances of each particular soil. The principal object is to produce heavy crops, which can only be secured by an almost exclusive use of the stronger-growing grasses and clovers.

Reverting to the question 'Is there no Alternative?' referred to at the opening of this chapter, it is interesting to remember that subsequent to the publication of a former edition of this work there appeared in 'The Times' an instructive communication from the late Earl of Leicester, K.G., describing the results of trials made at Holkham, Norfolk, with the object of securing, on poor soils, pasture that should last for a period of not less than six years. The mixture of seeds that was found to give the best results consisted of Cocksfoot, Perennial Rye Grass, Italian Rye Grass, Meadow Fescue, Hard Fescue, Tall Fescue, Timothy, Tall Oat Grass, Yellow Oat Grass, Alsike Clover, White Clover, and Yarrow. It is very desirable that the pasture should not be too closely grazed by sheep during the summer months of the first two or three years, and it is better, when practicable, to mow the crop the first year after laying down. The rye grasses ensure a sufficiency of herbage during the time other varieties are attaining full development. The two oat grasses are included because they are native to the Holkham district. With suitable modifications in the constituents, and a considerable addition to the quantity of seed per acre, as may be needed by local conditions of soil and climate, this mixture, which has given such good results on poor soils in Norfolk, might prove useful in other districts.

Temporary pastures are, as a rule, sown with spring corn, and they require substantially the same treatment as permanent pastures, although, as the grasses employed for the former purpose are strong growers, there is not quite the same necessity for extreme caution in preparing the land; but even here laxity and carelessness entail a sufficiently heavy penalty.

A temporary pasture may, in a favourable summer, afford a valuable bite for horned stock soon after the corn is cut; and as the grasses are robust and comparatively coarse, although none the less nourishing on that account, they will not be injured by the hoofs of cattle. The rolling should be done in November, instead of waiting until spring. Supposing stock to be kept off the ground, and the autumn prove warm and genial, it will sometimes be possible by the end of October to get a cut of useful green food.

The pasture needs bush-harrowing and rolling down early in spring before being laid in for hay. The first year's crop will mainly consist of Rye Grasses and Clovers, but the bottom of a three or four years' ley may be expected to improve for at least two years, and the Foxtail, Timothy, Cocksfoot, and other plants will increase in bulk in the third and fourth seasons.

A heavy dressing of cake-fed farmyard manure, applied towards the close of the year, will pay well. Nothing improves artificial grasses so much, and there must be no niggardliness in its use. The fresher the manure when placed on the ground, and the less it has been allowed to ferment, the better the grass will thrive. In the Kidmore grass experiments good results were also obtained from a dressing per acre of one hundredweight of nitrate of soda, with three-quarters of a hundredweight of muriate of potash, given in alternate years.

Leys which are intended to stand for a period of six or eight years are especially suited for poor or hilly land which cannot be profitably cultivated until an improvement has been effected in the texture of the soil. By a judicious selection of deep-rooted plants much can be done to increase the fertility of the land, reduce the cost of tillage, and discourage the growth of weeds. Among the plants which are adapted for the purpose are Burnet, Chicory, Kidney Vetch, Cocksfoot, Tall Fescue, Tall Oat Grass, Rib Grass, and Yarrow. Nearly all endure prolonged drought almost with impunity. The selection of these plants and the proportions of the seeding must be determined by the character of the field to be laid down. From a mixture wisely chosen paying crops for grazing or cutting are produced during the entire period of the ley. When the ground is broken up a crop of roots may be taken, followed by corn, and a similar course repeated for another term. Land which has long been regarded as of little value may thus be gradually brought into fertile condition, until it is amenable to the regular course system of the farm.

Hints on Haymaking

A GOOD deal has been written for and against the practice of mowing and grazing alternately. Strictly speaking, meadow land is always mown, and pasture land is always grazed. Although the terms and the practice are sometimes regarded as convertible, there is more in the distinction than appears at first sight. Certain grasses are better adapted for the scythe than for being eaten down by stock, and some fields contain a preponderance of one or more of these varieties. A meadow which answers to its strict definition should consist of those grasses which flower almost simultaneously, so that the entire crop may be ready for mowing at one time. Such a crop must, of necessity, be most unsatisfactory for grazing. There will be no early or late grasses for the cattle, but a heavy weight in June and July, which cannot be fed off economically. On the other hand, a good pasture, consisting of the grasses which ensure a continued succession of food, yields but a poor hay crop. However, the question now under consideration is not the best means of creating either the one or the other, but the turning into hay of grass set apart for that purpose.

The presence of stones, crocks, and other hard rubbish deposited by top dressings not only diminishes the crop while growing, but will by-and-by make it necessary to set the machine high enough to avoid breakage. Now the mower should always be set as low as possible, for the bottom herbage is, weight for weight, more valuable than the top, and every inch of the former counts both in weight and in quality. The clearing of stones from the field is therefore worth scrupulous attention, and it should be done before the grass makes a start in early spring. The turf must also be well bush-harrowed and firmly rolled down.

The time for mowing varies in different localities and in different seasons. But there are sound reasons for urging the importance of cutting the grass young, before even the earliest varieties have formed seeds in their flower-heads. In most grasses, and in all clovers, the formation of saccharine matter takes place in their stems during the early stage of growth, and the best hay is usually made from grass before the flowering heads have begun to turn colour. Experiments made in the chemical laboratory prove that, although there are exceptions, the great majority of grasses contain nearly double the quantity of nutritive matter before, than they do after, ripening seeds. This applies also to the clovers which form so large a proportion of every good meadow.

An objection to the early cutting of grass deserves a passing remark. It is quite true that young grass shrinks more than grass of older growth. In other words, a larger quantity of moisture is evaporated by the former, but as the loss is pure water only, it is of no importance whatever. Hay from an early mowing has the advantage over that which is cut later of being higher in quality and far more digestible, to say nothing of increased aftermath and the benefit conferred on the pasture by early cutting.[1]

[1] The following free extracts are from 'The Relative Feeding Value of Grass Cut at Different Periods of Growth,' by Martin J. Sutton and Dr. J. A. Voelcker, published in the *Journal of the Bath and West and Southern Counties Society*, Vol. II., Fourth Series, and subsequently issued by the Society as a separate pamphlet:—

'An old pasture at Kidmore was divided into five plots resembling each other in texture, aspect, surface soil, subsoil, and in the botanical constituents of the turf. These five plots were cut on the different dates named in the subjoined table. The total weight of hay produced for the season was practically the same for Plots Nos. 1 and 3; but the fact must not be lost sight of that the relative dryness of hay in the field is solely dependent on the state of the atmosphere at hay-time, and it is quite impossible, in a series of cuttings extending over the whole season, to take up the hay in the same degree of dryness each time. The real basis of comparison is not the weight of hay, nor the weight of green produce, but the dry weight after moisture

Mowing machines have greatly altered the conditions of haymaking, and the change is not always in favour of the hay. The temptation is to cut more grass than can be dealt with, and in wet seasons this may involve serious loss; for in a scorching time grass becomes hay almost without any making, and if not promptly ricked it may be so burnt up as to render the fibres hard and woody.

The stems of grass are protected by a thin coating of silicate, which has been termed 'Nature's waterproof mantle.' Tossing the grass about breaks the stems, and at the fractures moisture enters and decay is hastened. Hence in continued wet weather the cut grass should be allowed to lie just as it is left by the scythe or mower, when it will take the minimum of harm. In fine weather the tedding machine should be used much more freely than is commonly the case, for every time the machine goes over the ground a different surface is exposed to the sun. Often, when the end of a field is reached, the

had been expelled by a temperature of 212° Fahr. A glance at the table will show that the *total dry weight* of herbage comes out in the same order as the numbering of the plots—*i.e.* Plot No. 1 yielding the largest quantity, and No. 5 the smallest.

'By comparison of Plot No. 1 and Plot No. 4 (the first cutting of the latter Plot taking place simultaneously with the haymaking of the district) the loss resulting rom late mowing is only too apparent.

TOTAL CONSTITUENTS IN LBS. PER ACRE OF THE DIFFERENT CUTTINGS PRODUCED ON EACH PLOT.

—	Plot 1.* Cut June 3, Sept. 2, Oct. 29	Plot 2. Cut June 17, Sept. 17	Plot 3. Cut July 2, Sept. 29	Plot 4. Cut July 15, Oct. 29	Plot 5. Cut July 29, Oct. 29
	lbs.	lbs.	lbs.	lbs.	lbs.
Soluble Albuminoids	124	120	85	68	177
Insoluble Albuminoids	706	536	502	538	402
Soluble Carbo-hydrates, &c.	1,144	1,228	1,121	1,089	1,195
Digestible Fibre	2,461	2,294	2,104	1,947	1,874
Mineral Matter (excluding Silica)	442	368	370	376	348
Indigestible Fibre	1,860	2,008	1,897	1,755	1,769
Water	21,593	16,549	16,550	13,111	14,991
Silica	335	165	156	219	188
Total produce per acre, in lbs.	28,665	23,268	22,785	19,103	20,944
Total dry weight per acre, in lbs.	7,072	6,719	6,235	5,992	5,953'

* Only No. 1 Plot produced a third crop.

beginning is ready for turning again, and it is a mistake to suppose that because scorched grass makes bad hay, therefore quickly made hay resembles scorched grass. These remarks are intended to apply exclusively to meadow hay. Clovers, Sainfoin, &c., should be turned in the swath by hand, and with the utmost care, to avoid breaking off the leaves.

When a particular field is ready, the whole strength of the farm should be concentrated on the labour of gathering and carting the hay to the rick. I have known a crop of grass cut one morning and stacked at night; but the crop was light, the heat unusual, and the desiccation of the plant had considerably advanced before the mower was used. Although it is seldom done, I am persuaded that in hot weather grass might often with advantage be put into windrows on the day it is cut. Dew is well-nigh as injurious to half-made hay as rain, and grass which has parted with much of its water on a hot summer day is in a condition to reabsorb moisture from the atmosphere at night. This process goes on much more rapidly when the hay lies scattered on the ground than when it is raked together. The cocks should not be opened too early in the morning, and if the sun prove hot it will spoil the colour to scatter the grass very much. Greenness is one of the indications of well-made hay, while a brown shade, whether resulting from rain or sunburning, is a certain sign of deteriorated condition. Three days ought to make good hay in fair weather from an ordinary crop. Grass which is cut one day, tedded repeatedly the next, cocked that night, and opened out again on the following morning, may be fit to carry in the afternoon of that—the third—day. A very heavy crop, however, or a crop in which there is an unusual proportion of clover, must not be ricked so quickly, nor must it be left too thin on the ground. Succulent grass with large solid stems, full of moisture, is least easy to turn into hay, and is most liable to fire when ricked. This danger is often increased by fine weather, because the leaves and smaller shoots become ready to carry more quickly than the succulent stems. Those who have had experience with water-meadows are aware of the extreme hazard of carrying hay from them too soon. An old and safe test of fitness is to gather together a few of the stout stems and twist them tightly into a rope. If moisture exudes, the grass is not ready for the rick. Clover stems, Cow Grass especially, contain a large amount of moisture, and if ricked too soon there is a risk of the heat rising injuriously. Where clover forms a large proportion of the crop the haymaking machine should not be used.

Dry and benty grass which does not contain much clover will almost 'make itself' in fine weather, and but little fear need be entertained that it will fire. Indeed, if the rick does not heat somewhat the hay will be of inferior quality.

Ricks may occasionally be seen standing on the bare ground where floods are not unknown. It is unwise to erect them on such spots, for they absorb moisture almost as readily as a sponge. Some injury will certainly ensue from the damp ground, and the whole bulk of hay may be made mouldy by a flood. The cost of stone or iron staddles will often be repaid in a single year, and they render it safe to put hay into a rick much earlier than where this means of bottom ventilation is lacking.

The best shape for a rick is square. A thin oblong form exposes too large a surface to the atmosphere, and a very high-pitched roof is objectionable on the same ground. The bottom of the stack should be smaller than the eaves, so that rain falling from the thatch will clear the sides of the rick. In building, the centre should always be kept higher than the outside, and every layer of grass must be firmly trodden down. The roof must be steep enough to shoot off rain and snow, but it injures the upper portion of the hay to go higher than is necessary to secure this object. Roughly stated, the top of the rick measured across under the eaves should be about one and a half or two feet more than from the eave to the ridge. This gives almost an equilateral triangle.

Rick-cloths cost money, but they are of great service in protecting partially-made ricks when hay-carting is suddenly arrested by rain. They also render it easy to proceed with several ricks simultaneously, and therefore the grass can be stacked in a more moist condition than when a rick has to be hurried up and promptly thatched. Indeed, rick-cloths save all anxiety about thatching, for the task can be deferred until after the first rise of heat has been dissipated, and it is consequently safe to put the roof on.

The practice of making ventilating shafts in the centre of a rick by pulling up a box or sackful of hay as the work proceeds need only be resorted to in catching weather, when it is impossible to put the hay together in first-rate order. No doubt the contrivance has prevented many ricks from firing, but it restricts the partial fermentation which takes place in a solid stack, and this is an important influence in the manufacture of a fragrant sample of hay. A much better method of treating grass about the dryness of which there is a doubt, is to place layers of dry straw between thicker layers of hay. The straw will absorb the excessive moisture, and with it many of the valuable juices of the clovers, which will greatly enhance the feeding value. This procedure is not available for hay which is intended to be sold, but it will make capital fodder for home consumption.

'Weathered' hay—that is, hay which has been repeatedly soaked and dried until much of its value has been lost—will be improved in quality and made more palatable to stock by sprinkling coarse salt over the layers of the rick as the building proceeds. From ten to twenty pounds of salt should be used for each ton of hay.

After grass is cut for hay, it parts with nearly three-fourths of its weight by evaporation; but, except under the influence of long-continued rain, no chemical change of importance occurs in the field. In the rick, however, very considerable chemical changes take place, such as the creation of sugar by the action of heat on the starch contained in the grass. The difference between good and bad hay nearly as often results from too little or too great heat being evolved in the stack, as from faults in the process before stacking. Over-heating, even when it does not go so far as to blacken and char the hay, produces so much acetic acid as to make the fodder sour and unpalatable. Dr. Thompson showed that $387\frac{1}{2}$ pounds of grass are required to make 100 pounds of hay. The loss is chiefly water, but not entirely so. This is demonstrated by the fact that an animal which thrives on 100 pounds ot grass will not do nearly so well on 25 pounds of hay supplemented with 75 pounds of water. The loss of nutritious ingredients is of course attributable to the process of fermentation carried on in the stack. The sugar has been largely converted into alcohol and carbonic acid, by which a certain amount of waste has occurred.

However closely a field may be raked after the hay cart, a quantity of loose hay will remain scattered about, and it will be well worth while to turn in cattle to pick this up, and also to browse on the patches round the headlands and ditches which have escaped the mowing machine. Much waste is incurred by carelessness in this matter, and if the grass by the hedges and ditches is not eaten down while green, it will be unsightly all the summer. Immediately the cattle have consumed the fodder they must be driven out of the field, to give the aftermath time to make a fair start.

The making of aftermath hay is not by any means general, and is always precarious. Autumnal dews and shortening days, combined with the succulent nature of the herbage, are all against it. But as many of the late-growing grasses, of which the herbage principally consists, are specially nutritious, there is a prevalent opinion that the lattermath is of higher quality than the first cut.

The old Welsh system of 'fogging,' or allowing the hay crop to cure itself, uncut, in the open fields, demands but brief notice here. The fields are fed off by winter stock early in the spring up to the end of April; then the grass is laid in as if for hay, but is left uncut all the summer, and stock are not permitted to graze until December. Welsh farmers state that cows feeding on the dry grass give as good butter as in the spring of the year; that the frost makes the grass sweet; and that the herbage shoots very much earlier in the following spring, because it is protected by the brown growth of the previous year; also that when mixed with the old grass it is much more wholesome and sustaining food. The inducement to adopt this system is that all expense and anxiety of hay-making are avoided. Some farmers actually let their grass fields stand untouched from May until February or March of the following year, when the stock are turned out of doors. No doubt the practice supplies a great deal of food at a time of scarcity, and this food is specially prized for the early-calving cows. The whole system is, of course, contrary to all the recognised canons by which grass land is managed in England.

The quantity of hay annually produced in Great Britain differs widely from year to year, and this variation is mainly attributable to the character of the season. On the following page the Returns of the Board of Agriculture are quoted for ten years, from 1900 to 1909 inclusive. From these figures it will be observed that although the acreage does not greatly vary during the decade, there is an enormous divergence in the crop realised in different years. The smallest yield was 132,365,975 cwt. in 1901, and the largest weight 208,584,680 cwt. in 1907.

The estimated average production per acre of rotation hay in the former year is 25·48 cwt., and in the latter year 32·97 cwt., while from permanent grass the return is 16·63 cwt. in 1901 and 27·23 cwt. in 1907. Roughly speaking, it may be said that about two thirds of the annual production is obtained from permanent meadows and one third from clover, sainfoin, and grasses under rotation.

Year.	Areas. Acres.	Estimated Total Produce. Cwt.
1900	6,574,836	170,554,892
1901	6,706,517	132,365,975
1902	6,944,805	201,386,651
1903	7,166,720	195,061,446
1904	7,087,445	187,455,180
1905	6,876,998	164,627,200
1906	6,975,447	171,690,660
1907	7,186,254	208,584,680
1908	7,181,303	194,402,780
1909	6,812,467	167,369,020

Hints on Grazing

Most watchful care is needed to obtain the best results from grazing land, and only those who are actually engaged in the business have any adequate conception of the ceaseless anxiety it involves. Two main objects must be kept steadily in view—the constant progress of the stock, and the maintenance of fertility in the pasture. It may be necessary to change the cattle from field to field, or to alter the relative proportions of sheep and horned stock at very brief intervals. However arduous the labour, the cattle must be kept growing; and the grass be so fed off as to avoid waste and ensure continuous herbage. Allowance has also to be made for differences in seasons. In dry summers, what little grass there may be is extremely valuable for its high quality and sustaining power; but when vegetation is rank and sodden with moisture, a much larger quantity may fail to put on flesh. Truly 'the farmer's eye makes the beast fat.'

No precise date can be given for beginning to graze pastures in spring. Cattle should not be turned out until there is enough feed to keep them going without too much help from hay, nor until the ground is firm enough to prevent their hoofs from damaging the young shoots of the grasses. On the other hand, if the grass gets too old the animals will refuse much of it, and the fodder will be lost. Pastures consisting largely of early, strong-growing grasses, particularly Cocksfoot, will need to be stocked before others which produce finer and later varieties. Even after cattle have made a start, late spring frosts or a persistent east wind will sometimes upset the grazier's calculations, and the stock may begin to go back through scarcity of food. Then a supply from the Mangel clamp, or from a few acres of Cabbage or Thousand-headed Kale, will prove a boon indeed.

When sheep begin to graze in mixed pastures they will probably keep the plant down close; but, as herbage grows more rapidly, young horned stock should be added, so as to feed down the rising culms and flower-stalks which the sheep pass by. Sometimes neither sheep nor bullocks will eat the stems of Cocksfoot, and then it is necessary to run the mower over the pasture to prevent a copious seeding of this grass, which, however valuable in suitable proportions, is on certain soils a great pest when allowed to predominate. I have known cattle to derive benefit from grazing in one pasture by day and in a different field at night. It is interesting to note the eagerness which stock soon display for this change in their run.

A pasture specially adapted for fattening bullocks should not, as a rule, have sheep put on it; and a perfect sheep pasture should never have the grass long enough to feed a bullock; but from an ordinary pasture, devoted to mixed stock, the aggregate produce will be most profitably utilised by a judicious combination of horned stock, horses, and sheep. The grazier affords clear evidence of his skill by the manner in which he takes advantage of the special characteristics of every separate field as the seasons vary. In so proportioning stock as to feed the crop down evenly he benefits the pasture, and by changing the animals from field to field a difference of diet is ensured, for scarcely any two pastures produce exactly the same herbage. Each animal has its own likes and dislikes, but between them they are almost certain to clear the crop. Calves need the best grass that can be given them, and they generally thrive better alone than in company with other animals. Cows and sheep feed better together than do fat bullocks and sheep, and this mixture of stock is exceedingly good for the grass. But a first-rate bullock pasture is not certain to be a suitable place for milch cows; they are more likely to put on fat than to increase the flow of milk.

As fat beasts approach fitness for the butcher they become very dainty, and will often leave behind a lot of feed. This should be got rid of quickly by crowding the field with store stock. Then a brief rest will prepare it for bullocks again. It is a wise rule to start the stock on the worst pasture of the farm, and gradually move them forward to that which is better. Stock which have once had good food will not take kindly to that which is inferior, but when placed upon it they lose flesh, and this is ruinous for the grazier.

No matter how rich a pasture may be, I believe it to be sound practice for the beasts, for the grass, and for the farmer's pocket, to supply a liberal addition of Cabbage, Mangel, and above all of oil cake, before the feed runs at all short. A good grazier will not be content unless the stock lie down on the pastures a great part of every day. If they do not quickly satisfy appetite they cannot put on fat, and this is impossible when the animals are wholly dependent on grass which is eaten down close.

A stringent rule cannot be laid down as to when grazing land should be cleared of cattle for the winter. It must be done in time to ensure ample pasturage in the following spring, and the grass should be eaten down close before very cold weather sets in. It is surprising with what relish cattle will take some of the coarse grasses late in autumn which they have refused to eat during summer. Frost no doubt sweetens and makes these coarse tufts palatable. If they are not consumed they must be mown before Christmas.

The necessity of scattering the droppings of cattle is well understood, although it is often neglected. By postponing this work until late in the year a quantity of grass is lost to the grazier. Large unsightly patches of herbage appear, which the animals will not touch. Yet, after the grass of these patches is mown and has become withered, it is often sought for by animals affected with scour, and it has been stated to have a medicinal effect in curing them. The right thing to do, however, is to have the droppings scattered frequently, and in any case of neglect to have the tufts mown. An appliance attached to the chain harrow will effect the most complete distribution of all droppings.

A supply of pure water is a matter of great consequence for the animals. It is true they will drink that which is impure, stagnant, and filthy, but it is a vitiated taste, and such water is distinctly injurious to them. Tube wells often prove cheaper and better than the old system of pond-making.

Shade and shelter are also helpful to cattle. Trees and hedgerows save them from fierce sunshine, and ditches should not intervene to prevent access to the shade. Groups of trees in the field not only add to the beauty of the landscape, but prove beneficial to the animals on scorching days. Rough sheds, open in front and facing south, are desirable in early spring, as they afford protection from biting east winds and cold, violent rains. In these sheds lumps of rock salt should always be kept within reach of the stock.

Ensilage

(GRASS)

The preservation of green fodder, in an undried state, for winter feeding is not a recent discovery. The long and severe winters which prevail in Canada compelled the stock-keepers of that country to adopt the system long before it became familiar to farmers in the United Kingdom.[1] At first an underground silo, or a special construction above ground, was considered imperative, but it was soon discovered that air could be excluded from forage stored in old barns and outhouses, and silage has even been made in casks. It is therefore obvious that costly buildings are not essential, and the erection of them has practically ceased. The silage stack is generally preferred. By this contrivance sweet or sour silage can be made at will, and the atmosphere can be effectually prevented from injuring more than a few inches of the exterior of the fodder. The stack also offers an advantage over the fixed silo, as it can be erected on any spot that will save haulage and facilitate feeding.

Subtropical plants such as Maize and Sorghum are grown expressly for conversion into silage, and almost every kind of herbage which this country produces has been experimented on, with more or less satisfactory results. Here I am only alluding to ensilage in relation to grass, and, so far as its application to this crop is concerned, I am persuaded that the use of the silo or silage stack will, in favourable

[1] In so different a climate as that of the islands of the South Seas the natives avail themselves of the principle of the silo for the preservation of bread-fruits.

seasons, be the exception. Grass will continue to be turned into hay very much as heretofore. When a farmer can convert his grass into hay in three genial days, it is improbable that he will consent to cart nearly four times the weight of green fodder to the silo, with the risk of failure in the end.

Still it has been demonstrated that in wet summers the silo or silage stack is sometimes a real boon. In districts where the average of seasons is unfavourable to haymaking, ensilage has materially modified the conditions of profitable farming. The reports of greatest success come from farms worked under the alternate system of three or four years' leys. Prescriptions of grasses, clovers, &c., specially adapted for growing suitable herbage for the silo produce the best silage.

Lattermath grass may with advantage be sent to the silo or silage stack, on account of the difficulty of making hay in autumn, and, as the late crop has the reputation of possessing more 'proof' than the summer cutting, it will be all the more valuable for ensilage. The decision to convert a crop of grass into silage should be acted upon from the outset. It is a mistake to suppose that hay badly washed by rain can yield good silage. In wet seasons the hay is not only more costly to make, but when made is of low feeding quality; so that silage in such years will not only be cheaper but superior.

Those who have tried silage as food for dairy cows are practically unanimous in its favour, although there is some risk in feeding it alone. In fattening bullocks, however, the Rothamsted experiments prove that, for putting on flesh, swedes and mangels have a considerable advantage over grass or clover silage, whether sweet or sour.

Analyses show that in the process of conversion there is a serious loss in the nutritive qualities of green provender after it is placed in the silo.

Notes on Certain Worthless Grasses and Injurious and Poisonous Plants found in Pastures

No feature of modern farming is more calculated to excite a feeling of astonishment than the indifference shown to the presence of worthless grasses and pernicious weeds in pastures. On too many farms something green, and plenty of it, appears to be the supreme object of attainment. Whether the crop is made into hay, or beasts and sheep are turned out to eat what they can, and reject or tread down the remainder, the animals are expected to thrive on the herbage.

Every arable farmer knows that unless land is kept clear of couch, black grass, thistles, &c., the crops of corn must suffer ; but it does not appear to be equally evident to the grazier that certain plants, however useful they may be elsewhere, must in a pasture be classed among weeds. It is no sufficient answer to say that the plants increase the bulk of herbage. They are none the less weeds on that account. The objectionable character of some other plants is freely admitted, and feeble, intermittent attempts may possibly be made to eradicate those which are easily recognised, and are admitted to be directly injurious or actually poisonous to sheep, and especially to milch cows. The erroneous impression that it is difficult or impossible to eliminate these worthless or injurious plants from grass land has deterred many farmers from attempting the task, or, at all events, there has not been the persistence that would have achieved success. Were it only possible to deal with the pests of grass land by the plough and scarifier, as on arable soil, the effort would be hopeless.

Happily, the weeds of pastures are amenable to treatment which, though far less costly, is more effective, lasting, and remunerative in its results. Some few of the deep-rooted plants, as thistles, coltsfoot and rushes, need repeated attention with the scythe or spud for two years; but the majority of pasture weeds succumb to well-regulated and suitably adjusted doses of chemical manures, which beneficially stimulate the nutritious grasses. The grazier may therefore clean and enrich his land at the same time, and prove by his own experience that as a rule the poor pasture is foul, while the rich pasture is clean.

To effect an improvement takes time and needs patience. A pasture or meadow which has deteriorated through years of neglect cannot be won back to first-class condition in a single season, nor can a good and clean crop be maintained in its most profitable state without unceasing vigilance. It is, however, reassuring to know that the necessary cost of labour, and the outlay for manures which favour the development of the stronger-growing grasses and at the same time discourage the growth of many forms of pernicious vegetation, will on pastures eventually yield a far more abundant return than can be recouped from a similar expenditure on arable land.

In order to prevent the deterioration of pastures, or to improve the herbage, worthless plants should be promptly recognised, and methods of eradication be understood. This involves a knowledge of the pests of pastures, their duration, mode of perpetuation, and other peculiarities which may suggest the readiest means of compassing their destruction. Large weeds, such as thistles, docks, and coltsfoot, are too conspicuous to escape notice. Less obtrusive but equally baneful subjects are too often allowed to live on; and the bad grasses, because they happen to be grasses, continue to flourish, to the exclusion of species rich in feeding constituents.

As a rule, the low condition of a pasture is most manifest near its boundaries. This may arise from the neglect of hedgerows and ditches, or from the carelessness of neighbours, who permit the dissemination of weed seeds, to their own loss and that of adjoining owners. In this matter it is not often possible to control the action of others, but it is the duty of every occupier to destroy with an unsparing hand the evil tenants of his own land.

From the series of experiments made by the late Sir J. B. Lawes and Sir J. H. Gilbert at Rothamsted, and confirmed by my own independent trials at Dyson's Wood and Kidmore Grange, it has been demonstrated that most of the annual and not a few of the perennial weeds can be, for all practical purposes, eliminated from a pasture by applications of combined mineral and ammoniacal salts. Judicious dressings of these salts augment the growth of the stronger grasses, so that buttercups, dandelions, plantains, sorrel, and *Briza media* are starved out, and even *Bromus mollis* is considerably reduced in vigour. Lime alone is a powerful aid in promoting luxuriance among the better class of herbage, and is also distinctly prejudicial to weeds, especially to those of annual duration. Early cutting, too, has produced remarkable effects in discouraging weed life. In the years 1891 and 1892 part of a meadow at Kidmore Grange, thickly strewn with weeds, was cut in advance of the usual haymaking season, and in the second year a marked diminution of worthless plants was perceptible, as compared with the remaining portion, which had been cut at the time usual in the district.

To enumerate all the worthless plants found in pastures is beyond the design of this chapter. In alluding to those which are most common and injurious to stock, opportunity will be seized to indicate the readiest means of reducing their numbers and of preventing their perpetuation.

Coltsfoot (*Tussilago Farfara*).—Pastures on poor, stiff clays are not infrequently overrun with this weed. The creeping underground stems render spudding impossible, but the yellow flower-heads, which appear in advance of the leaves during February and March, should be cut, to prevent seed being ripened; and another cutting when the leaves are about half-grown will weaken the plants.

Couch, or Twitch (*Triticum repens*).—A pest that is universally known, and appears to be almost irrepressible. The rootstocks are stout, and minute portions will grow and spread in all directions, robbing the land of its fertility. On light soils which have not been thoroughly cleaned before sowing seeds it is very troublesome, but as the turf becomes consolidated the couch diminishes in quantity, and under good management it eventually disappears. It is never found in rich old pastures.

Docks (*Rumex sp.*) are found on all farms. The species most troublesome in grass lands are *Rumex obtusifolius* and *Rumex crispus,* both perennials, flowering and seeding abundantly. In the South of England, the flowering period is about the first or second week in June, and the destruction of plants should be accomplished before that time. By means of the docking iron, when the ground has been softened by rain, the whole of the tap-root can be removed, and it is important not to leave a remnant, or fresh growth will give renewed trouble. Should there be no opportunity of drawing the roots, frequent cutting at the collar will weaken the docks, and when persistently followed up will reduce their numbers. It is a mischievous practice to throw docks into ditches or under hedges; very little moisture is requisite to maintain life and enable them to perfect seed.

Comparatively few samples of clover harvested in this country are entirely free from dock seed, but it is true economy to ensure a pure sample.

Foxtail, Slender (*Alopecurus agrestis*).—This annual, the Black Twitch of the Midlands, also called Hungerweed, is chiefly troublesome in temporary pastures on sandy and other light soils, and the objection to it is that the mass of root-fibres absorbs the available plant-food in the soil, without making a return in wholesome herbage. Consolidation of the turf is unfavourable to its existence, and in grass kept close and firm it is starved out. The seed of Slender Foxtail, sometimes in considerable proportion, is found in low-priced samples of *Alopecurus pratensis*.

Knapweed, or Hardheads (*Centaurea nigra*).—A tough, much-branched perennial, common to many pastures and leys, on both wet and dry soils. Where it is very abundant or shows a tendency to spread, the use of ammonia salts and mixed minerals will almost banish this weed from grass lands.

Ragwort (*Senecio Jacobæa*).—A coarse, perennial weed, allied to the common groundsel, sometimes attaining a height of four feet. It throws up bright golden flower-heads, seeds freely, and is common in the light-soil pastures of many districts. Cattle avoid it, but sheep are said to eat the leaves when young. A single plant will spread over a space eighteen inches in diameter. The roots are brittle, and it is futile to attempt to draw them in spring, as every scrap left in the ground will produce a vigorous plant in the following season. Cutting the flower-heads before full development results in the shrinking and hardening of the roots, which are then easily drawn when the soil has been softened by autumnal rains.

Rib Grass, or Narrow-leaved Plantain (*Plantago lanceolata*).—A free-seeding, deep-rooting perennial, common to most soils. Its presence among grass is countenanced by some farmers, who have yet to be convinced that other and more nutritious herbage might profitably occupy its place; on this point comparative analyses leave no room for doubt. Whatever manurial dressings are favourable to the grass crop are prejudicial to Rib Grass. The seed is one of the most frequent impurities of Red Clover, and Perennial Rye Grass often contains a considerable percentage of it.

Sedges and Rushes (*Carex sp. and Juncus sp.*).—Even under vigilant supervision these plants will sometimes intrude on the low-lying portions of water-meadows, and if neglected they multiply so rapidly as practically to exclude the gramineous herbage. The most effectual mode of destroying these semi-aquatics is to drain the land thoroughly; where this is impossible the herbage must be cut low several times in every season.

Soft Brome Grass (*Bromus mollis*).—The early and abundant production of seed facilitates the spread of this pernicious grass. So rapidly is the seed-crop developed that the leaves and stems are drained of the little nutrition they possessed, and the herbage is therefore worthless as hay. As Soft Brome Grass is of annual, or at most of biennial, duration, early mowing for two or three seasons reduces it in meadow lands, and in pastures it can be diminished by a dressing of ammonia salts combined with mineral manures.

Sorrel, Common (*Rumex Acetosa*), is closely allied to docks, and is found in nearly all meadows, sometimes in great profusion. The plant sends down its tap-root to a considerable depth, and destruction by hand is almost impracticable. Dressings of mineral and ammonia salts weaken sorrel, and at the same time assist the grass crop.

Thistle, Creeping (*Carduus arvensis*).—When once established in a pasture this weed is difficult to extirpate. Soils rich in lime are its favourite home. On such land the strong rootstocks creep in all directions, and penetrate to a depth of two or three feet. Constant spudding in early spring, when the shoots are full of sap, will hasten the exhaustion of this perennial weed.

Thistle, Marsh (*Carduus palustris*), often abounds in low-lying marshes, and sends up a solitary erect stem to a height of three or four feet. A large quantity of seed is ripened, and disseminated over a wide area, during July and August. Spudding low down, before the flower-heads have fully developed, is the best means of destroying this annual thistle.

Tufted Hair Grass, or Tussock Grass (*Aira cæspitosa*).—The great raised tufts of this grass which are seen in some low-lying meadows are a proof of neglect and mismanagement. Its herbage is worthless, and is so harsh and dense as to interfere with both scythe and machine at hay-time. The roots impoverish the land, and where a few plants have been allowed to develop, they should be removed by grubbing. Efficient draining will always ensure the extirpation of Tussock Grass. The seed of this grass is often one of the impurities of *Alopecurus pratensis*, and of some other grasses used in laying down land.

Woolly Soft Grass, or Yorkshire Fog (*Holcus lanatus*).—One of the most persistent and troublesome of the worthless grasses, possessing structural endowments which ensure its continued existence under extremely varied conditions. The perennial roots form a dense mat near the surface, and the plant endures both drought and frost almost with impunity. The flowering period extends from June until late in autumn, and if left undisturbed a large quantity of seed is perfected. Manuring appears to have little effect in reducing the vitality of Woolly Soft Grass, unless a high degree of fertility is continuously maintained. Early mowing tends to weaken the plant, and prevents the seed from maturing. Depasturing and the close treading of land by sheep have been known to lessen the proportion of this grass; and it is certainly advisable, in the event of flower-heads showing in autumn, to skim the meadows with a scythe.

Seed of *Holcus lanatus* is found more or less in all inferior samples of *Alopecurus pratensis*, *Cynosurus cristatus*, and some other grasses used in prescriptions for sowing down permanent and temporary pastures.

Yellow Rattle (*Rhinanthus Crista-galli*).—A parasitic annual which preys on the roots of grasses and clovers in many water-meadows and poor, damp pastures. Cattle dislike it, and as the seeds are matured before meadows are usually mown, there is full opportunity for its increase. Salt at the rate of three or four hundredweight per acre, applied to those portions of the pasture where Yellow Rattle abounds, will greatly reduce and may entirely exterminate the plant.

The plants already described are negatively objectionable, being deficient in those qualities which are essential in crops intended for the support of animal life. Economic principles demand that such plants should be destroyed, and their places filled with herbage which contributes to the credit side of the agricultural balance-sheet.

The plants to which I have now to refer possess the far more serious fault of containing active poisons, or of being injurious in some other way, such as tainting milk and rendering butter unsaleable. Cattle, and especially in-calf cows, frequently show a predilection for unusual herbage, and the death of valuable animals may be the first intimation that watercourses, hedges, and ditches have not been kept free from poisonous plants ; or the loss of a good market for milk and butter may be almost equally disastrous. When cattle are first turned out of their winter quarters, tufts of Garlic-Mustard, Yellow Rocket, and Crow Garlic are very tempting; but they are all milk-tainting plants, and should not be permitted to exist where live-stock have access.

Buttercups (*Ranunculus acris and R. bulbosus*).—These two varieties are common in meadows and pastures, and in their green state taint the milk of cows that eat them. The volatile, acrid constituent is dissipated when the plants are dried and made into hay. Of the two, *R. bulbosus* is less acrid than *R. acris*, and stock are less disposed to eat the latter, which flowers later in the season than the former.

A satisfactory mode of extirpating Buttercups has yet to be discovered. They are too numerous to be dealt with individually, but they occupy so much space that no reasonable effort should be spared to weaken their growth. Early mowing checks them, and a generous dressing of farmyard and artificial manures assists the valuable herbage, while it discourages the development of Buttercups.

Crowfoot, Celery-leaved (*Ranunculus sceleratus*), is by far the most poisonous of our native Ranunculi. It immediately affects the milk of cows that eat the leaves, and has been known to cause the death of young stock. The habit of the plant is erect; the leaves glossy, rather fleshy, and divided, somewhat resembling those of celery; the small pale yellow flowers appear throughout the summer, and, like the leaves, emit a strong nauseous odour when bruised. This species can be exterminated by pulling up the plants, which is very easily done; and if ditches and ponds are thoroughly cleaned for a couple of seasons, during May or June, Celery-leaved Crowfoot will cease to be troublesome.

Deadly Nightshade (*Atropa Belladonna*).—A poisonous plant, found in waste places on chalky and limestone soils, which is easily distinguished by its solitary purple flowers and the fetid odour of the leaves when bruised. Whenever this plant is discovered it should be effectually destroyed.

Fetid Mayweed, or Stinking Chamomile (*Anthemis Cotula*).—A white-flowered, annual, compositaceous plant, having furrowed stems and finely-cut leaves covered with minute resinous glands, secreting an acrid principle which blisters the skin when the plant is handled. The offensive odour differentiates this from other Mayweeds, such as the Common Chamomile and the Corn Chamomile. The species is common in rough pasture lands and leys, and should be kept under by the repeated use of a sharp hoe.

Foxglove (*Digitalis purpurea*), **Tansy** (*Tanacetum vulgare*), and **Wormwood** (*Artemisia Absinthium*) are well known to be injurious plants, and should not be allowed to exist within reach of cattle and sheep.

Garlic, Crow (*Allium vineale*).—The green, fistulous, onion-like leaves of this plant show conspicuously in pastures and meadow land early in spring, while the grass is yet short, and the tufts appear to be very tempting to milch cows, whose milk is at once affected by its pungent, acrid properties. The flowers, or rather the heads of bulbils, are thrown up in July, and if allowed to mature they disperse and grow with great rapidity. Persistent hand-pulling of the bulbous roots early in the year is the only way of ridding pastures of Crow Garlic.

Garlic-Mustard (*Alliaria officinalis* or *Sisymbrium Alliaria*).—A rank annual cruciferous plant, which emits the odour and tastes strongly of Garlic. It is common in hedge-rows, and early in the spring its alluring bright green leaves are often eaten by cows, when a very unpleasant flavour is imparted to milk and butter. The boundaries of pastures should be freed from this plant before the white flower-heads appear in May. Seed is freely perfected in June or July. This plant is also known by such names as Hedge Mustard, Sauce Alone, and Jack-by-the-hedge.

Hemlock, Common (*Conium maculatum*).—A strong-growing biennial, found in hedgerows and woodsides of dry and sandy soils. In the second year it grows to a height of four or five feet, and flowers in July. The leaves are bright green, very large, and cut into fine segments; the stem large, hollow, and spotted with purple. The plant possesses in a marked degree the qualities of a narcotic poison, and when bruised it emits a powerful mouse-like odour which is very characteristic. By cutting off the rank growth early in the season flowering and seeding may be prevented; and as this is one of our most injurious native plants, continuous efforts should be made to destroy it.

Hemlock, or Water Dropwort (*Œnanthe crocata*).—A coarse-growing, umbelliferous, poisonous perennial, attaining a height of three or four feet, and having deep green compound leaves. The umbels of white flowers appear in July. In the South of England and the Midlands this plant often abounds in ditches and by the side of watercourses to which cattle have access. Cutting down the plants in spring and early summer, and keeping ditches, &c., clean, are the most efficacious means of preventing mischief from this dangerous plant.

Meadow Saffron (*Colchicum autumnale*).—A poisonous bulbous plant, which is not uncommon in light-soil pastures and meadows in the Midlands. The pale purple flowers appear in autumn, followed by broad, glossy, lily-like leaves in the spring. The bulbs should be forked up immediately the plant is discovered.

Water Pepper (*Polygonum Hydropiper*).—An evil-smelling, pungent annual, which grows in ditches and watercourses, and affects the milk of cows if eaten. It should be rooted out and destroyed.

Yellow Rocket (*Barbarea vulgaris*).—A yellow-flowered cruciferous plant, which often grows in abundance by ditches and streams, and in early spring throws up large tufts of turnip-like leaves, which, notwithstanding a pungent, nauseous flavour, are freely eaten by cattle, with the result that milk and butter are tainted. Before stock are turned on to the pastures in spring a labourer should be sent round with a hoe to cut the tufts below the crown, and as the plant is of biennial duration only, it can be easily destroyed. Seeding should not be allowed.

As a rule, the wise and safe course is to collect all injurious plants after they are cut or uprooted, and burn them.

Spottiswoode & Co. Ltd., Printers, London, Colchester and Eton.

Educational Cabinets—*continued.*

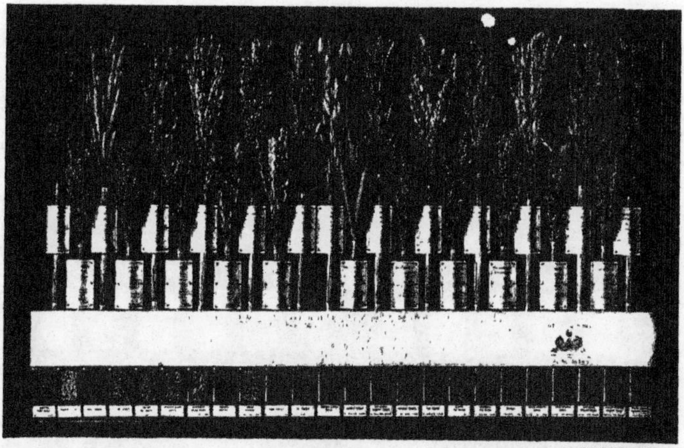

COPYRIGHT S & S

No. 2.—Polished Oak, glazed, Educational Cabinet, with sun-blind, containing 24 varieties of Natural Grasses suitable for Permanent and Temporary Pastures, giving popular and botanical names, soils and localities where usually found, with analyses by Dr. Voelcker of the Grasses, both dried and green, and 24 glazed boxes of seeds of the same Grasses. Size, 34 in. by 22 in. Price **20s.**

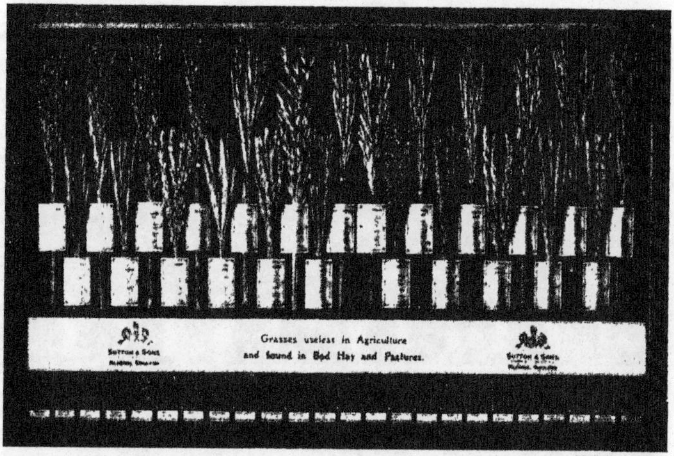

COPYRIGHT S. & S

No. 3.—Polished Oak, glazed, Educational Cabinet, with sun-blind, containing 24 varieties of Grasses *useless* in Agriculture, and found in bad hay and pastures, giving popular and botanical names, soils and localities where usually found, with 24 glazed boxes of seeds of the same Grasses. Size, 34 in. by 22 in. Price **20s.**

Educational Cabinets—*continued.*

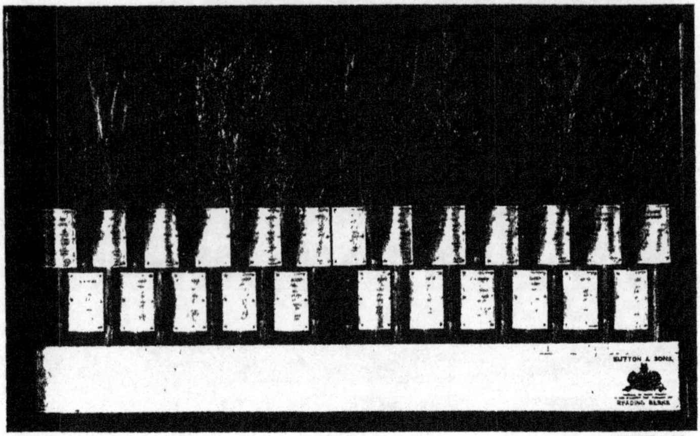

COPYRIGHT S. & S.

No. 4.—Glazed Educational Cabinet, containing 24 varieties of Natural Grasses suitable for Permanent and Temporary Pastures, giving popular and botanical names, soils and localities where usually found, with analyses by Dr. Voelcker of the Grasses, both dried and green. Size, 28 in. by 17 in. Price **8s. 6d.**

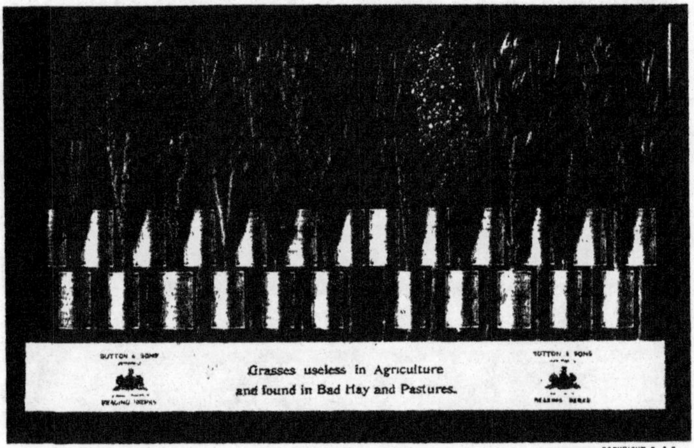

COPYRIGHT S. & S.

No. 5.—Glazed Educational Cabinet, containing 24 varieties of Grasses *useless* in Agriculture, and found in bad hay and pastures, with popular and botanical names, soils and localities where usually found. Size, 28 in. by 17 in. Price **8s. 6d.**

Educational Cabinets—*continued.*

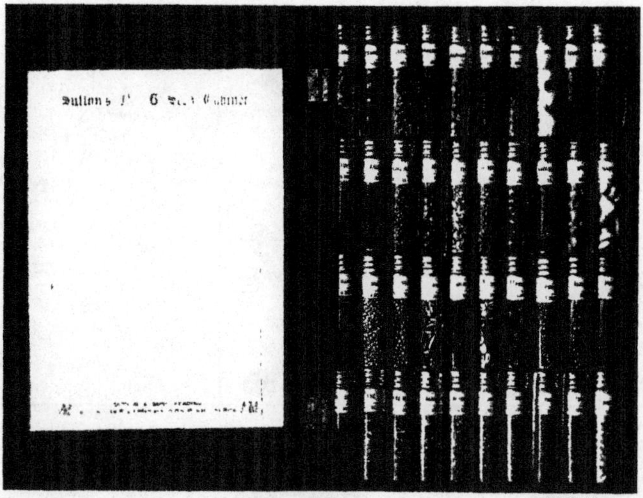

COPYRIGHT S. & S.

No. 6.—Wooden Case, containing 40 varieties of Agricultural and Horticultural seeds in metal-capped bottles, giving popular names, with a table showing the use of each variety. Price **8s. 6d.** The Set of 3 (Nos. 4, 5, 6), **25s.**

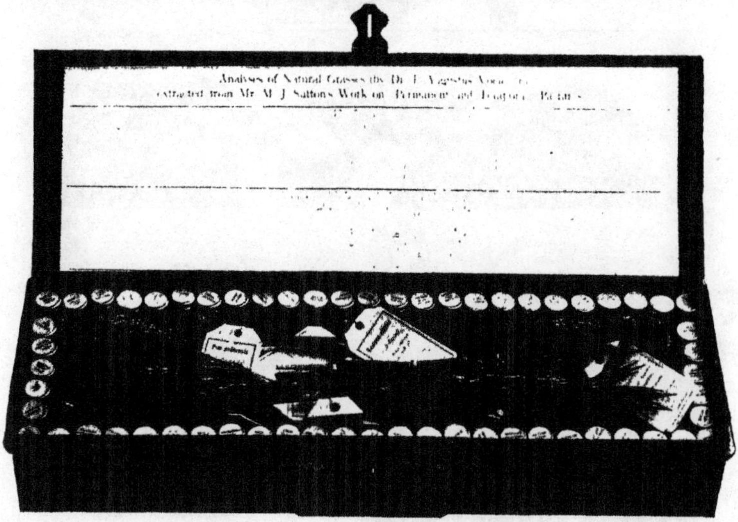

COPYRIGHT S. & S.

No. 7.—Sutton's Composite Collection of Useful and Useless Grasses, in Japanned Vasculum, with brass handle and fittings, containing 36 varieties of Grasses, giving popular and botanical names, soils and localities where usually found, with analyses by Dr. Voelcker of the Useful Grasses, both dried and green; also 60 metal-capped glass bottles of Grass and Agricultural seeds. Price **10s. 6d.**

Grasses, Clovers, Seeds, &c.

COPYRIGHT S & S

No. 8. — Botanical specimens of Natural Grasses, Leguminous Plants, and certain Weeds (see list on following page), with popular and botanical names. Mounted on cardboard, **1s.** each ; Unmounted, **6d.** each.

No. 9. — Grasses (both useful and useless in Agriculture) made up in bundles, with popular and botanical names, soils and localities where usually found.
Price **2s. 6d.** per dozen.

COPYRIGHT S & S

No. 10.—Grass seeds and Agricultural seeds, in metal-capped bottles.
Per dozen, **3s.**

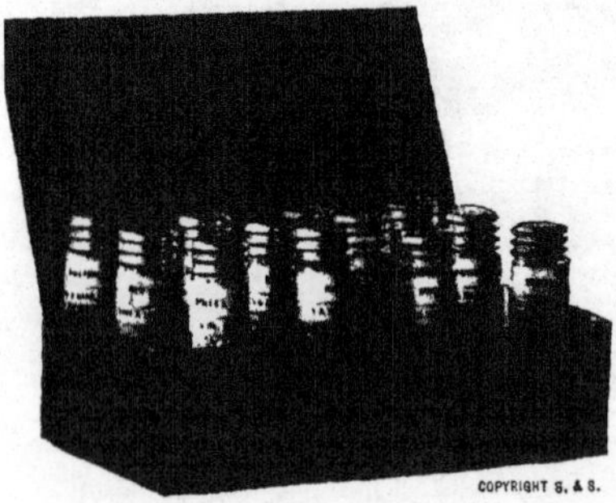

COPYRIGHT S. & S.

COPYRIGHT S. & S.

No. 11.—Weed seeds, of the commoner sorts (see list, page viii), in small metal-capped bottles.
Per dozen, **5s.**

List of Weed Plants referred to under No. 8.

AGRIMONY
(Agrimonia Eupatoria).
AVENS
(Geum urbanum).
BEDSTRAW, GREAT HEDGE
(Galium Mollugo).
BEDSTRAW, LADY'S
(Galium verum).
BINDWEED, SMALL
(Convolvulus arvensis).
BROOM-RAPE
(Orobanche minor).
BUCKWHEAT, CLIMBING
(Polygonum Convolvulus).
BUGLOSS, SMALL
(Lycopsis arvensis).
BUGLOSS, VIPER'S
(Echium vulgare).
BURDOCK
(Arctium Lappa).
BUTTERCUP, MEADOW
(Ranunculus acris).
BUTTERCUP, BULBOUS
(Ranunculus bulbosus).
CAMPION, WHITE
(Lychnis vespertina).
CAMPION, BLADDER
(Silene inflata).
CARROT, WILD
(Daucus Carota).
CAT'S-EAR
(Hypochæris radicata).
CHARLOCK
(Sinapis arvensis).
CHICKWEED
(Stellaria media).
CHICKWEED, MOUSE-EAR
(Cerastium vulgatum).
CLEAVERS
(Galium Aparine).
COLTSFOOT
(Tussilago Farfara).
CORNFLOWER
(Centaurea Cyanus).
CRESS, WINTER
(Barbarea vulgaris).
CROWFOOT, CELERY-LEAVED
(Ranunculus sceleratus)
CROW GARLIC
(Allium vineale).
DAISY, COMMON
(Bellis perennis).
DAISY, OX-EYE
(Chrysanthemum Leucanthemum).
DANDELION
(Taraxacum officinale).
DOCK, LARGE
(Rumex obtusifolius).
DROPWORT, HEMLOCK
(Œnanthe crocata)
DYER'S GREEN WEED
(Genista tinctoria).
DYER'S WEED
(Reseda luteola).
FUMITORY
(Fumaria officinalis).
GOAT'S BEARD
(Tragopogon pratensis).
GOOSEFOOT
(Chenopodium album).
GROUNDSEL
(Senecio vulgaris).
HAREBELL
(Campanula rotundifolia).
HEMLOCK
(Conium maculatum).
HEMP NETTLE
(Galeopsis Tetrahit)
HENBANE
(Hyoscyamus niger).
HORSETAIL
(Equisetum arvense).
KNAPWEED
(Centaurea nigra).
KNOTGRASS
(Polygonum aviculare).
LADY'S SMOCK
(Cardamine pratensis).
MADDER, FIELD
(Sherardia arvensis).
MALLOW
(Malva sylvestris).
MANGEL, WILD
(Beta maritima).
MARIGOLD, MARSH
(Caltha palustris).
MARJORAM, WILD
(Origanum vulgare).
MAYWEED, SCENTLESS
(Matricaria inodora).
MAYWEED, STINKING
(Anthemis Cotula).
MEADOW VETCHLING
(Lathyrus pratensis).
MELILOT, COMMON
(Melilotus officinalis).
MIGNONETTE
(Reseda lutea).
MUGWORT
(Artemisia vulgaris).
MUSTARD, GARLIC
(Sisymbrium Alliaria).
MUSTARD, HEDGE
(Sisymbrium officinale).
NETTLE, STINGING
(Urtica sp.).
NIGHTSHADE, BLACK
(Solanum nigrum).
NIGHTSHADE, DEADLY
(Atropa Belladonna).
NIGHTSHADE, WOODY
(Solanum Dulcamara).
NIPPLEWORT
(Lapsana communis).
PANSY, WILD
(Viola tricolor).
PARSLEY, FOOL'S
(Æthusa Cynapium).
PARSNIP, WILD
(Pastinaca sativa).
PEARLWORT
(Sagina procumbens).
PERSICARIA, COMMON
(Polygonum Persicaria).
PERSICARIA, BITING
(Polygonum hydropiper).
PLANTAIN, BUCKSHORN
(Plantago Coronopus).
PLANTAIN, GREATER
(Plantago major).
PLANTAIN, HOARY
(Plantago media).
PLANTAIN, RIBWORT
(Plantago lanceolata).
POPPY, CORN
(Papaver Rhœas).
RADISH, WILD
(Raphanus Raphanistrum).
RAGWORT
(Senecio Jacobæa).
RATTLE, YELLOW
(Rhinanthus Crista-Galli).
REST HARROW
(Ononis arvensis).
ST. JOHN'S WORT
(Hypericum perforatum).
SCABIOUS, FIELD
(Scabiosa arvensis).
SELF-HEAL
(Prunella vulgaris).
SHEPHERD'S PURSE
(Capsella Bursa-pastoris).
SILVER WEED
(Potentilla anserina).
SNAPDRAGON
(Linaria vulgaris).
SORREL, COMMON
(Rumex Acetosa).
SORREL, SHEEP'S
(Rumex Acetosella).
SORREL, WOOD
(Oxalis Acetosella).
SPEEDWELL, IVY-LEAVED
(Veronica hederæfolia).
TANSY
(Tanacetum vulgare).
TEAZLE
(Dipsacus sylvestris).
THISTLE, LARGE
(Carduus sp.)
TREFOIL, HARE'S-FOOT
(Trifolium arvense).
VETCH, TUFTED
(Vicia Cracca).
WOODRUFF
(Asperula odorata).
WOODRUSH, FIELD
(Luzula campestris).

List of Weed Seeds referred to under No. 11.

AGRIMONY
(Agrimonia Eupatoria).
AVENS
(Geum urbanum).
BEDSTRAW, GREAT HEDGE
(Galium Mollugo).
BEDSTRAW, LADY'S
(Galium verum).
BINDWEED, LARGE
(Convolvulus sepium).
BINDWEED, SMALL
(Convolvulus arvensis).
BLUEBELL
(Scilla nutans).
BROOM-RAPE
(Orobanche minor).
BUCKWHEAT, CLIMBING
(Polygonum Convolvulus).
BUGLOSS, SMALL
(Lycopsis arvensis).
BUGLOSS, VIPER'S
(Echium vulgare).
BURDOCK
(Arctium Lappa).
BUTTERCUP
(Ranunculus sp.).
CAMPION, WHITE
(Lychnis vespertina).
CAMPION, BLADDER
(Silene inflata).
CARROT, WILD
(Daucus Carota).
CAT'S-EAR
(Hypochæris radicata)
CHARLOCK
(Sinapis arvensis).
CHICKWEED
(Stellaria media).
CLEAVERS
(Galium Aparine).
COLTSFOOT
(Tussilago Farfara).
CORN COCKLE
(Agrostemma Githago).
CORNFLOWER
(Centaurea Cyanus).
CRESS, WINTER
(Barbarea vulgaris).
CROWFOOT, CELERY-LEAVED
(Ranunculus sceleratus).
CROW GARLIC
(Allium vineale).
CUCKOO PINT
(Arum maculatum).
DAISY, COMMON
(Bellis perennis).
DAISY, OX-EYE
(Chrysanthemum Leucanthemum).
DANDELION
(Taraxacum officinale).
DOCK, LARGE
(Rumex obtusifolius).
DOCK, SMALL
(Rumex crispus).
DODDER
(Cuscuta Trifolii).
DROPWORT, HEMLOCK
(Œnanthe crocata).
DYER'S GREEN WEED
(Genista tinctoria).
DYER'S WEED
(Reseda luteola).
FOXGLOVE
(Digitalis purpurea).
FUMITORY
(Fumaria officinalis).
GOAT'S BEARD
(Tragopogon pratensis).
GOOSEFOOT
(Chenopodium album).
GROUNDSEL
(Senecio vulgaris).
HEMLOCK
(Conium maculatum)
HEMP NETTLE
(Galeopsis Tetrahit)
HENBANE
(Hyoscyamus niger).
KNAPWEED, GREAT
(Centaurea Scabiosa).
KNAPWEED
(Centaurea nigra).
KNOTGRASS
(Polygonum aviculare).
MADDER, FIELD
(Sherardia arvensis).
MALLOW
(Malva sylvestris).
MANGEL, WILD
(Beta maritima).
MARIGOLD, CORN
(Chrysanthemum segetum).
MARIGOLD, MARSH
(Caltha palustris).
MAYWEED, SCENTLESS
(Matricaria inodora).
MAYWEED, STINKING
(Anthemis Cotula).
MEADOW SWEET
(Spiræa Ulmaria).
MELILOT, COMMON
(Melilotus officinalis).
MIGNONETTE
(Reseda lutea).
MUGWORT
(Artemisia vulgaris).
MUSTARD, GARLIC
(Sisymbrium Alliaria).
MUSTARD, HEDGE
(Sisymbrium officinale).
NETTLE, STINGING
(Urtica sp.).
NIGHTSHADE, BLACK
(Solanum nigrum).
NIGHTSHADE, DEADLY
(Atropa Belladonna).
NIPPLEWORT
(Lapsana communis).
PANSY, WILD
(Viola tricolor).
PARSLEY, BUR
(Caucalis sp.).
PARSLEY, FOOL'S
(Æthusa Cynapium).
PARSNIP, WILD
(Pastinaca sativa).
PENNYCRESS
(Thlaspi arvense).
PEPPERWORT
(Lepidium campestre).
PERSICARIA, COMMON
(Polygonum Persicaria).
PERSICARIA, BITING
(Polygonum hydropiper).
PILEWORT
(Ranunculus Ficaria).
PLANTAIN, BUCKSHORN
(Plantago Coronopus).
PLANTAIN, GREATER
(Plantago major).
PLANTAIN, HOARY
(Plantago media).
PLANTAIN, RIBWORT
(Plantago lanceolata).
POPPY, CORN
(Papaver Rhœas).
RADISH, WILD
(Raphanus Raphanistrum).
RAGWORT
(Senecio Jacobæa).
RATTLE, YELLOW
(Rhinanthus Crista-Galli).
ROBIN
(Geranium dissectum).
ST. JOHN'S WORT
(Hypericum perforatum).
SCABIOUS, FIELD
(Scabiosa arvensis).
SELF-HEAL
(Prunella vulgaris).
SHEPHERD'S PURSE
(Capsella Bursa-pastoris).
SORREL, COMMON
(Rumex Acetosa).
SORREL, SHEEP'S
(Rumex Acetosella).
SPEEDWELL, IVY-LEAVED
(Veronica hederæfolia).
TEAZLE
(Dipsacus sylvestris).
THISTLE, LARGE
(Carduus sp.).
VETCH, TUFTED
(Vicia Cracca).
WOODRUFF
(Asperula odorata).
WOODRUSH, FIELD
(Luzula campestris).

GARDEN LAWNS, TENNIS LAWNS, BOWLING GREENS, CROQUET GROUNDS, PUTTING GREENS AND CRICKET GROUNDS.

By SUTTON AND SONS.

COPYRIGHT S. & S.

CONTENTS.

Introduction—Soil—Drainage—Grass Slopes—Preparatory Work—Weed Seeds in Soil—Enriching the Soil—Surface Preparation—Selection of Seeds—Quantity of Seed—Sowing—Bird Scares—Worm Casts—Watering the Seed Bed—Mowing—Rolling—Destruction of Weeds—Annual Top-dressing—Improving Lawns—Fairy Rings—Moss in Lawns—Plans of Tennis Court and Croquet Ground—Putting Greens—Cricket Grounds.

BEAUTIFULLY ILLUSTRATED. **PRICE ONE SHILLING.**

'The information bearing on the practical work of laying down new lawns and replenishing and maintaining old turf will be helpful to amateurs and professionals alike. The peculiarities and functions of the different varieties of grasses are briefly described, and the particulars given as to the quantity of seed to sow, the time and manner of sowing, the preparation of the soil, drainage, and so forth are compiled from intimate knowledge of the subject in hand.'—**THE TIMES.**

'There is practically nothing to do with grass growing that is not fully dealt with in this publication, whether it be beautiful garden lawns and slopes or lawns for such purposes as tennis, croquet, and bowls. Cricket grounds and putting greens are also subjects treated from the expert's point of view.'—**LADIES' FIELD.**

Through all Booksellers, from

Simpkin, Marshall, Hamilton, Kent & Co. Limited, London

or direct from

SUTTON AND SONS, READING.

The illustration above shows the method of selecting roots for the production of Sutton's pedigree stock of Prizewinner Yellow Globe Mangel. Every year several tons of typical specimens are chosen. The roots are separately examined and laid out in rows for final inspection by one of the partners, whose experience enables him to detect the smallest departure from the true type. Conformity to an ideal is sought in every root. There must be freedom from side roots, the true colour, and, above all, the density which indicates high feeding quality. The same process is adopted with Swedes, Turnips, Cabbage, and other farm crops offered by the Firm as a result of more than a century's effort to make farming profitable.

It is the **constant selection** of each variety that has gradually established the value of Sutton's pedigree stocks, which prove their superiority by winning premier acreage prizes at leading competitions in all parts of the Kingdom.

Many lists contain the offer of seeds bearing the name of Sutton's, but none are genuine without the registered trade mark which appears on a ticket enclosed in every sack or bag from the Warehouses of the Royal Seed Establishment, Reading.

REGISTERED TRADE MARK.

Printed in Dunstable, United Kingdom

91220734R00097